SPECIAL PUBLICATIONS
THE MUSEUM
TEXAS TECH UNIVERSITY

Catalogue of Type Specimens of Neotropical Bats in Selected European Museums

Dilford C. Carter and Patricia G. Dolan

No. 15 July 1978

TEXAS TECH UNIVERSITY

Cecil Mackey, President

The Museum
Special Publications No. 15
136 pp.
28 July 1978
$8.00

Special Publications of The Museum are numbered separately and published on an irregular basis under the auspices of the Dean of the Graduate School and Director of Academic Publications, and in cooperation with the International Center for Arid and Semi-Arid Land Studies. Copies may be obtained on an exchange basis from, or purchased through, the Exchange Librarian, Texas Tech University, Lubbock, Texas 79409.

ISSN 0149-1768
ISBN 0-89672-063-2

Texas Tech Press, Lubbock, Texas

1978

CONTENTS

Catalogue of Type Specimens of Neotropical Bats in Selected European Museums

Dilford C. Carter and Patricia G. Dolan

This catalogue is the outgrowth of a project begun by D. C. Carter in 1966. A concerted effort was made to locate, examine, and measure type specimens of tropical American bats in eleven of the principal European museums. The institutions visited included the major repositories for collections resulting from the zoological explorations of the New World by Europeans between the end of the eighteenth century and the beginning of the twentieth. It was on these collections that so many of the Neotropical taxa of bats now recognized were based.

Although most European museums have endeavored to label old type specimens, the task has been a difficult one because mammalogists through the first seven decades of the 1800's viewed species as having a constancy that did not require type specimens or precise locality data. Prior to the time Oldfield Thomas began his long and productive tenure at the British Museum, authors of species descriptions seldom recorded more than the vaguest reference to the specimen(s) on which they based a name, even though a few (especially Peters, at the Berlin museum) often wrote lengthy, detailed descriptions. Localities typically were defined in the most general terms, often the country of origin being deemed sufficient even when an author knew more precisely where a specimen in hand had come from (for example, see Peters, 1856:415, and our account of *Vampyrus auritus* Peters). Except for a series of papers begun by Rode (1938) listing type specimens of mammals at the Muséum National d'Histoire Naturelle, Paris, there are no recent catalogues of type specimens in European museums comparable to those of Poole and Schantz (1942) for the United States National Museum or Goodwin (1953) for the American Museum of Natural History. In the process of compiling our list of primary type specimens, we have attempted to provide the user of this catalogue with accurate bibliographic references, precise locality and field data (within the limitations that control such a task), descriptive detail where we thought it would be of taxonomic value, and pertinent remarks of historical or nomenclative import.

METHODS

Four months of 1966 were spent examining specimens of Neotropical bats in the museums listed below; the time devoted to each collection was allocated according to the number of specimens contained therein. For the purpose of this catalogue, the British Museum was revisited in 1976. Abbreviations preceding the names of institutions are used (with museum catalogue numbers) in the accounts beyond to identify specimens.

BMNH British Museum (Natural History), London
MNCN Museo Nacional de Ciencias Naturales, Madrid
MNHN Muséum National d'Histoire Naturelle, Paris
NMW   Naturhistorisches Museum Wien, Vienna
NR   Naturhistoriska Riksmuseum, Stockholm
RNH   Rijksmuseum van Natuurlijke Historie, Leiden
SMF   Natur-Museum und Forschungs-Institut Senckenberg, Frankfurt a.M.
SMNS Staatliches Museum für Naturkunde in Stuttgart, Stuttgart
ZMB   Zoologisches Museum der Humboldt-Universität zu Berlin, Berlin
ZMH   Zoologisches Staatsinstitut und Zoologisches Museum, Hamburg
ZSM   Zoologisches Staats-Sammlung München, Munich

In addition to examining all specimens labeled as types, a search was made of each collection for previously unrecognized primary typological material; essentially, all specimens in a collection were examined. Museum catalogues, specimen labels (including curatorial notations), receipts, invoices, and other pertinent museum records, in addition to original descriptions and subsequent taxonomic works, were used to corroborate the typological status of the specimens listed in this catalogue. With respect to curatorial notations, we attempted to determine the author as well as the relevance. Specimen invoices, such as the one written by Joseph Natterer at Vienna and annotated by Wagner at Munich, were extremely helpful; those written by dealers in natural history objects (as well as their labels) were essentially useless because a specimen's "origin" too often seems to have been determined by a museum's desiderata.

Format

In the accounts of type specimens, families, subfamilies, and genera are arranged in systematic order following Simpson (1945), Cabrera (1958), and Hall and Kelson (1959). Mormoopidae, not recognized as a family in any of the preceding three works, is placed between the Noctilionidae and Phyllostomatidae following Smith (1972). The arrangement of phyllostomatids is based on that of Jones and Carter (1976). Species group names are arranged alphabetically within the currently recognized genus to which they belong; in most cases, our generic assignments of species do not differ from those of Cabrera (1958) or Hall and Kelson (1959); where they do, the reasons for reassignment are addressed in the remarks for that account.

Each account consists of the species group name, reference to the first published description or indication for the name, museum and catalogue number, age and sex (when it was possible to determine these), method of preservation, locality of origin, collector or source (including collector's field number and measurements if noted on the specimen label), date of capture (if known), type status, descriptive statements about a specimen, and remarks (generally of a historical or nomenclative nature) when we thought them appropriate. Additional statements concerning the components of a species account follow.

Species group name.—The name for each account is spelled and capitalized exactly as it was printed in the heading for the original description. Discrepancies in the spelling of a name within the original description are noted in the remarks.

Citation.—Reference to the first published description or indication for a name includes year of publication, title of work or abbreviated name of serial, volume or number (if any), page number on which the name first appeared in conjunction with its description (if the name of a new species group taxon appeared also on a page preceding the description, it is noted in the remarks), plate and figure numbers for those taxa illustrated, and day and month of publication when it was possible to determine this. It was sometimes impossible to ascertain with certainty a work's date of publication; the *Monatsberichte der Königlichen preussischen Akademie der Wissenschaften zu Berlin,* a monthly publication of papers presented at meetings of the Akademie, was one of the more troublesome series. The titles of works are spelled out but shortened if excessively long. The names of serials are abbreviated; although our system does not follow precisely that used in the List of Serial Publications in the British Museum (Natural History) Library, second edition, 1976 (for 1975), a reader should have no difficulty locating the name of an unfamiliar serial in the British Museum list.

Museum catalogue number.—Specimens are identified by museum (abbreviated) and catalogue number; "not numbered" indicates that none had been assigned at the time an institution was visited. Numbers separated by a solidus indicate that they were written originally in this fashion, that two numbers have been applied to the specimen, or that a skull was numbered separately from its skin or body. Occasionally, two or more numbers had been assigned independantly to a specimen; in such cases, we usually selected one (the earliest if we could determine the chronological sequence) and listed the others in the remarks section.

Age and sex.—Adults and subadults were distinguished on the basis of epiphyseal fusion with (or separation from) the shaft of metacarpals or on the degree to which the basioccipital was ankylosed to the basisphenoid. Sex was determined by the genitalia. Our statement that the sex was undetermined indicates that the genitalia had been removed in preparation. Discrepancies in sex recorded in an original description and our observation are noted in the remarks section.

Method of preparation.—Fluid-preserved specimens are referred to as being in alcohol; if the skull had been removed it was noted. Standard museum skins and skulls are identified as "skin and skull." Collectors in the early part of the nineteenth century often attempted to prepare lifelike mounted specimens of bats, with wings spread, for display in a museum's public gallery (the skull was not removed in this process, and the mouth was fixed in an open position so as to expose the teeth); examples preserved in this way are referred to as "skin, skull not removed," or "skin, skull removed" (the skulls of some were removed subsequent to their preparation).

Locality.—Localities were determined as accurately as possible by use of specimen labels, museum records, original and subsequent descriptions by the author of a name, revisions and other published works, and maps (both old and new). Whenever possible, a locality includes place name, department, province, or state, and country. Locality data enclosed in brackets are from a secondary source (those not involving the author of a name or collector); generally, we have used this method to add department, province, or state and country names to localities that were incomplete as originally written or reported. A citation is given for localities restricted or determined by someone other than ourselves. With few exceptions, place names are spelled in the official language of the country from which specimens came. Obvious errors, or minor variances, in spelling of place names have been corrected, or standardized, without notation. Quotation marks are used to identify ambiguous or questionable localities. Alternate spellings are given in brackets when we thought it appropriate to do so.

Collector (or source) and date of capture.—When not included on the specimen label, this information was obtained from museum records or published works. The origin of some specimens could not be traced to a collector, but for most of these we were able to determine the source (that is, the person or institution from which a museum obtained a specimen). Date of capture could not be ascertained for most specimens collected through the first eight decades of the nineteenth century, although Johann Natterer often recorded the date on his field labels.

Collector's measurements.—At the beginning of the twentieth century, collectors began to assign individual field numbers and to record four standard external measurements (length of head and body, tail, hind foot, and ear). This sequence of measurements is followed, without special notation, in the accounts beyond; additional measurements are identified.

Type status.—Generally, old primary type specimens in European museums are labeled simply as "type"; occasionally, specimens are labeled cotype (meaning syntype). We have identified specimens as "holotype or syntype" when we could not determine the number of specimens available to an author. We designated lectotypes for a few nominal species for reasons explained in the remarks.

Description.—Partly because the original purpose of this project was to obtain information of use to research then in progress of W. B. Davis and D. C. Carter, but also because some specimens are of greater nomenclatural importance than others, descriptive detail reported herein for skins and skulls is somewhat uneven among species. Unless stated otherwise, the reader may assume that a specimen is in reasonably good condition, that is, it can be handled with normal care and most measurements can be taken with confidence. The condition of a skull usually can be judged by the variates that we were able to measure. Selected specimens from the Texas Cooperative Wildlife Collection (Department of Wildlife and Fisheries Sciences, Texas A&M University, College Station, 77843) were used for comparative purposes; these are identified by the prefix TCWC plus museum catalogue number.

Remarks.—For the most part, comments in this section of the accounts are restricted to historical and nomenclative matters. Comments on specific (or subspecific) relationships or on the synonymy of names were considered to be extralimital to the purpose of this catalogue, except where our findings based solely on type specimens were in obvious contradiction to traditional views commonly held in 1977. Where appropriate, we have referred the reader to recent revisions.

AUTHORSHIP

Throughout this catalogue, we have maintained a strict interpretation of Article 50 in the International Code of Zoological Nomenclature, second edition, 1964, which is

> Article 50. Author of a name.—The author (authors) of a scientific name is (are) the person (persons) who first publish(es) it [III] in a way that satisfies the criteria of availability [IV], unless it is clear from the contents of the publication that only one (or some) of the joint authors, or some other person (or persons), is alone responsible both for the name and the conditions that make it available.

In the case of names (*Vespertilio caninus*, *V. nigricans*, *V. calcaratus*, and *V. leucogaster*) sometimes assigned by subsequent authors to Wied-Neuwied (*in* Schinz, 1821), Schinz is the author because Wied-Neuwied was not responsible for the conditions that made these available, even though he might have been responsible, in a sense, for the names. The fact that Wied-Neuwied (1826) later cited Schinz (1821) for those same names further indicates to us that Wied-Neuwied cannot be considered the author of these names by our current definition of authorship.

Wagner (1843) published descriptions of 19 species of Brazilian bats, specimens for 14 of which had been collected by Johann C. Natterer. Smith's (1977) argument notwithstanding, Wagner is the author of all 14 names attributed to Natterer. In those days, an author's assignment of a name he was describing to another person was not uncommon, and in Wagner's case does not imply that Natterer was responsible both for the names and the conditions that made them available. The names attributed to Natterer were more or less the same as those appearing in Natterer's diary (for example, Natterer's *Molossus gymnonotus* was changed to *Chilonycteris gymnonotus*), but Wagner's diagnoses were not taken from Natterer's field journal. Our photocopy of Natterer's original notes annotated by Wagner does not support Smith's contention that the descriptions were those of Natterer. See also the account for *Chilonycteris gymnonotus.*

EMBALLONURIDAE

Rhynchonycteris

Emballonura lineata Temminck

1840.   Monographies de Mammalogie, 2:297. (1835)

RNH 17642: adult of undetermined sex; skin, skull removed; Surinam; H. H. Dieperink; date of capture not specified. Lectotype (Husson, 1962:33).

Skin.—Banding on forearm indistinct.
Skull.—Occiput slightly damaged. First upper premolar small, laterally compressed.

Remarks.—This specimen is listed in Jentink (1888) as *Rhynchonycteris naso*, specimen *a*.

Proboscidea saxatilis Spix

1823.   Simiarum et vespertilionum brasiliensium species novae . . . , p. 62, pl. 35, fig. 8.

ZSM 22/50: adult of undetermined sex; skin, skull not removed; Brazil [Rio São Francisco (Burmeister, 1854:64)]; Spix; date of capture not specified. Syntype.

Skin.—Somewhat faded. Bands on forearm.
Skull.—Apparently entire.

ZMB 3085: adult female; in alcohol; Brazil [Rio São Francisco (Burmeister, 1854:64)]; received on exchange from the Zoologisches Staats-Samlung München; collector and date of capture not specified. Syntype.

Skin.—Somewhat faded.

RNH 17643: adult of undetermined sex; skin, skull not removed; Brazil [Rio São Francisco (Burmeister, 1854:64)]; Spix; date of capture not specified. Syntype.

Skin.—Bands on forearm.

Remarks.—All three specimens apparently are from Spix and rank equally as syntypes. An old label for ZSM 22/50 bears the notation "*Emballonura naso* Wied / *Proboscidia saxatilis* Spix / ?*Proboscidi rivalis*"; a label on the bottle that contains only ZMB 3085 indicates that this specimen is a type; and a specimen label with RNH 17643 identifies it also as a type. RNH 17643 was listed as *Rhynchonycteris naso*, specimen *b*, in Jentink's (1888) catalogue.

Saccopteryx

E[mballonura]. insignis Wagner

1855.   Die Säugthiere . . . von Schreber, Supplementband, 5:695.

NMW (not numbered): adult female; skin; "Registro do Sai," Brazil; Johann Natterer, no. 20; 26 April 1818. Holotype or syntype.

Skin.—Dorsal pelage dark, dull brown (flight membrane similar in color), two pale gray stripes on back; venter paler than dorsum.
Skull.—Not located and presumed lost.

Remarks.—An old label, probably the original, bears the notation "N 20. Registro do Sai. 26 Apr. 18. Foem." Wagner described this species as variety β of *Urocryptus bilineatus.*

Saccopteryx bilineata centralis Thomas

1904.   Ann. Mag. Nat. Hist., ser. 7, 13:251. April.

BMNH 88.8.8.20: adult female; in alcohol, skull removed; Teapa, Tabasco, México; collected by H. H. Smith, presented to the British Museum by O. Salvin and F. D. Godman; date of capture not specified. Holotype.

Skin.—Located but not examined.

Skull.—Rostrum inflated, flat dorsally; facial ridges indistinct; sagittal crest present; rostrofacial angle 150°; skull shaped similar to that of *Saccopteryx canescens*, but much larger; basisphenoid pit with low partition along midline.

Remarks.—Thomas examined "about a dozen specimens," but did not specify museum numbers or localities for these paratypes. Only the names Salvin and Godman appear on the skin label.

Saccopteryx canescens Thomas

1901.   Ann. Mag. Nat. Hist., ser. 7, 7:366. April.

BMNH 99.11.2.2: adult; probably female; skin and skull; Obidos, Amazonas, Brazil; J. Trumbull, ear measured 11.5; 21 May 1898. Holotype.

Skin.—Condition reasonably good for that of a small emballonurid. Dorsal color of pelage similar in pattern to that of *Rhynchonycteris naso*; venter cream, individual hairs with basal band gray, median band pale brown, distal band cream; forearms naked; membranes blackish, ears brownish.

Skull.—Zygomata broken, premaxillaries missing. Rostrum inflated; rostrofrontal angle 145°; facial ridges present from postorbital processes to sagittal crest; sagittal crest low; basisphenoid pit with low partition along midline.

Remarks.—The label notation for the ear measurement might not be that of the collector. The skin and skull of this specimen look very much like those of the holotype for *Saccopteryx pumila*, and we question the recognition of both *canescens* and *pumila*. Thomas referred to other specimens of S. *canescens* in the British Museum from Surinam and the Orinoco but gave no additional details; measurements were given for a male preserved in alcohol, but it was not identified by number or locality.

Saccopteryx gymnura Thomas

1901.   Ann. Mag. Nat. Hist., ser.7, 7:367. April.

BMNH 75.10.22.2: adult female; in alcohol, skull removed; Santarém [on the Rio Amazonas, Pará, Brazil]; Wickham; date of capture not specified. Holotype.

Skin.—Condition fair, some loss of pelage.

Skull.—Similar in size and shape to that of *Saccopteryx canescens* except that rostrum not so broad and maxillary toothrows not so far apart.

Remarks.—Apparently only a single specimen (*c* of *Saccopteryx leptura* in Dobson, 1878) was examined.

Saccopteryx pumila Thomas

1914.   Ann. Mag. Nat. Hist., ser. 8, 14:410. November.

BMNH 98.5.8.4: adult male; skin and skull; Altagracia, lower Río Orinoco, Venezuela; G. K. Cherrie, no. 9662; 12 January 1898. Holotype.

Skin.—Back of skin torn in preparation, some hair lost. Color as in *Rhynchonycteris naso*; dorsum apparently with two whitish stripes, as in *S. canescens* (traces of white visible); venter darker than in *canescens* (individual hairs with median band of reddish color darker and less distinct, distal band more buff than cream); forearms naked (therefore without whitish bands); membranes and ears blackish.

Skull.—Zygomata, postorbital processes, mandible, and part of maxilla broken. Like that of *Saccopteryx canescens*.

Remarks.—Thomas separated *pumila* from *canescens* on the basis of larger molars and the absence of a basisphenoid septum in *pumila*; otherwise, the holotypes of *Saccopteryx canescens* and *S. pumila* are so similar in appearance that we suspect a single species is represented by the two specimens. The description was based on four specimens, one taken by S. M. Klages at a locality near Lake Valencia, Carabobo, Venezuela, and one or two from Cayenne, French Guiana.

Urocryptus bilineatus Temminck

1838.   Tijdschr. Natuurl. Gesch. Physiol. Leiden, 5:33, pl. 2, figs. 3, 4.

RNH 17461: adult of undetermined sex; skin and skull; Surinam; H. H. Dieperink; date of capture not specified. Holotype.

Skin.—Condition poor.

Skull.—Damaged, extracted subsequent to receipt of the prepared skin by the Rijksmuseum.

Remarks.—RNH 17461 was listed in Jentink (1888) as *Saccopteryx bilineata*, specimen *a*.

Cormura

Emballonura brevirostris Wagner

1843.   Arch. Naturgesch., 9(1):367.

NMW (not numbered): adult male; in alcohol, skull removed (in alcohol with body); "Baraneiva" [Bananeira, Mato Grosso], Brazil; Johann Natterer; date of capture not specified. Syntype (but identified as cotype, meaning paratype, of *Myropteryx pullus* Miller).

Skin.—Faded, some loss of hair. Color now pale brownish; wing sac present, proximal to humerus-radius joint, anterior border of opening near anterior margin of antibrachial membrane, opening directed laterally to body.

Skull.—As in Miller's description of the genus *Cormura*, except that second upper premolar with cingulum (not mentioned by Miller, 1907:91), in contact with upper canine; basisphenoid pit not divided by longitudinal septum.

NMW (not numbered): adult female; in alcohol, head removed; "Baraneiva" [Bananeira, Mato Grosso], Brazil; Johann Natterer; date of capture not specified. Syntype.

Skin.—Condition poor, specimen falling apart.

Skull.—Head removed; skull lost.

Remarks.—The inscription "*Emballonura brevirostris* Wagn / Ex coll. Natt / 1844 / Bras," in Wagner's handwriting, appears on the face of the old green label attached to the second specimen listed above; inscribed on the back, "Type! Die inzige Species u. das inzix Examplar! Peters, N. Ak. W. Berlin 1867 / Schadel nicht auffindbar!" The original label was not with the specimen.

According to the invoice of specimens sent by Joseph Natterer to Wagner in 1843, the name *Emballonura brevirostris* was based on three specimens, all in spirits from an unspecified locality; Wagner's statement that the specimens were from Maribitanas appears to be an error caused by Joseph Natterer having listed two of the three specimens of *E. brevirostris* on a line beneath two specimens of *Emballonura canina*, one of which was from Maribitanas. Pelzeln (1883) stated that Johann Natterer took three specimens of *E. brevirostris* at Baraneiva, and that is the locality written on the specimen labels glued to the jars containing the two specimens to which this account refers. We were unable to find a place by that name although this locality must have been in western Mato Grosso because Natterer was in that region in 1829, a date that corresponds with one given for one of the specimens of *E. brevirostris* seen by Wagner. Johann Natterer left Villa Bella Cidade de Mato Grosso, (now known as Villa Bella de Santissima Trinidad, or simply as Mato Grosso, 15° 1′S, 59° 57′W) in July of 1829 and descended the Rio Guapore to the Rio Mamore to the Rio Madeira to the Rio Amazonas, arriving at Barra do Rio Negro in June 1830. Given the problem that early travelers had with the spelling of place names in sparcely populated areas of tropical America, we believe that a place now known as Bananeira (approximately 10° 39′S, 65° 23′W), on the Rio Mamore (there forming the boundary between Brazil and Bolivia), in the state of Mato Grosso, is most likely the place referred to by Johann Natterer as Baraneiva.

Myropteryx pullus Miller

1906.   Proc. Biol. Soc. Washington, 19:60. 1 May.

ZMB 3360: adult female; in alcohol, skull removed; Surinam; Kappler; date of capture not specified. Holotype.

Skin.—Pelage reddish brown.
Skull.—Not located and presumed lost.

Remarks.—The name *Myropteryx pullus* first appears on p. 59 as the type species for *Myropteryx*, which is described on that page, but the specific description begins on p. 60. Miller referred to four specimens, but only two of these, a paratype and the holotype, were found in the Berlin museum. Both are numbered 3360.

The specimen (adult female) in the Vienna museum that was sent by L. von Liburnau to Miller with the information that the "type of *Cormura*, originally in the Natural History Museum [Vienna]" could not be found, is one of the three specimens taken by Johann Natterer at "Baraneiva" and therefore is a syntype of *Emballonura brevirostris* Wagner (1843:367). Although the adult female in the Vienna museum is identified as a cotype of *M. pullus*, we have listed this specimen as a syntype of *E. brevirostris* (see account above).

In spite of differing opinions on the validity of *Myropteryx* and *Cormura* (see Sanborn, 1937; Cabrera, 1958), *Myropteryx* was based in part on a syntype for the name *Emballonura brevirostris*, the type species for *Cormura*.

Peropteryx

Emballonura macrotis Wagner

1843.   Arch. Naturgesch., 9(1):367.

NMW (not numbered): adult female; in alcohol, skull removed; [here restricted to Cuyabá (Cuiabá)], Mato Grosso, Brazil; Johann Natterer, no. 87; date of capture not specified. Syntype.

Skin.—Condition poor. Dorsal pelage dull, moderately dark reddish brown; ventral pelage paler.

Skull.—Not located and presumed lost.

Remarks.—The two labels (both old) in the specimen bottle are inscribed "*Emballonura macrotis*" and "XII Mato Grosso. (87.) adult." (both appear to have been written by Wagner). Wagner examined a second specimen (not located by us), which apparently came from Cuyabá, as pencilled on the Vienna loan invoice to Johann Wagner, dated 1843.

Peronymus cyclops Thomas

1924.   Ann. Mag. Nat. Hist., ser. 9, 13:531. May.

BMNH 24.3.1.6: adult male; skin and skull; Tushemo, near Masisea, 1000 ft., [on Río] Ucayali [in department of Loreto], Perú; Latham Rutter, no. 255, measured 50, 15, (8), 17; 3 October 1923. Holotype.

Skin.—Prepared rather poorly. Dorsal pelage dark reddish brown; ventral pelage buff brown (paler than on back); wings white distal to elbow, dark brown or blackish proximal to elbow; uropatagium dark brown or blackish; chiropatagium along side of body edged in white; ears dark brown or blackish, apparently not joined across top of head.

Skull.—Damaged and glued. Rostrum much inflated laterally (as broad as braincase), not so much inflated dorsally; facial-rostral angle 158°; facial ridges rather weakly developed, extend from postorbital processes to sagittal crest; basisphenoid pit apparently without median partition (there appears to be some glue in the pit).

Remarks.—Thomas examined a single specimen, taken by Latham Rutter; on the specimen label, the date appears to be 2 October, and the collector's name, misspelled (Rudder). Tushemo and Tushma might be variations in the spelling of the same place name. The altitude of Masisea and surrounding area is approximately 500 feet.

Peropteryx Kappleri Peters

1867.   Monatsb. Kön. preuss. Akad. Wiss. Berlin, p. 473.

ZMB 3348: adult female; in alcohol, skull removed; Surinam; Kappler; date of capture not specified. Holotype.

Skin.—Faded.

Skull.—Not located and presumed lost.

Remarks.—A single specimen was available to Peters. *Peropteryx leucoptera,* also collected by Kappler in Surinam and described by Peters (p. 474), was not located at the Berlin museum.

Centronycteris

Centronycteris centralis Thomas

1912.   Ann. Mag. Nat. Hist., ser. 8, 10:638. December.

BMNH 0.7.11.3: adult male; skin and skull; Bogava [Bugaba], 250 m., Chiriquí, Panamá; H. J. Watson, no. 31, measured 60, ——, 6, 15; 20 October 1898. Holotype.

Skin.—Dorsal pelage gray at base, buff yellow distally; ventral pelage paler than dorsal; membranes blackish; ears brownish.

Skull.—Damaged, occiput broken and missing, zygomata missing, postorbital processes and mandible broken. Rostrum inflated laterally; nasals short, dorsal emargination on border of exterior nares; facial ridges weak; sagittal crest developed; basisphenoid pit with well-developed median partition; first upper premolar with three cusps (one main cusp and well-developed anterior and posterior cingular styles); rostrum with longitudinal depression on dorsal surface.

Remarks.—Thomas distinguished this specimen from *Centronycteris maximiliana* by "its very different basisphenoid pits." The locality (Bogava) written on the skin label is an erroneous spelling for Bugaba (82° 29′N, 82° 37′W).

Vesp [*ertilio*]. *calcaratus* Schinz

1821.   Das Thierreich . . . , 1(Säugethiere und Vögel):180.

ZMB (not numbered): adult of undetermined sex; in alcohol; Brazil [Fazenda Coroaba, Rio Jucú, near Rio do Espírito Santo (Wied-Neuwied, 1826: 271)]; Maximilian, Prinz zu Wied-Neuwied; date of capture not specified. Holotype or syntype.

Skin.—Condition poor, originally a mounted skin. Ears relatively long and rather pointed.
Skull.—Partial skull, now in alcohol.

Remarks.—The specimen was collected by Wied-Neuwied and must be one of those on which the name *V. calcaratus* was based. Although the name has been attributed to Wied-Neuwied (*in* Schinz, 1821), Schinz is the author of the name. Subsequent descriptions by Wied-Neuwied (1822-31, 1826) cited Schinz (1821:180).

Balantiopteryx

Balantiopteryx io Thomas

1904.   Ann. Mag. Nat. Hist., ser. 7, 13:252. April.

BMNH 86.9.3.1: adult male; skull; Río Dolores, near Cobán, Alta Verapaz, Guatemala; R. C. Sarg; date of capture not specified. Holotype.

Skin.—Not located; body in alsohol when examined by Thomas.

Skin.—Not located; body in alcohol when examined by Thomas.

Remarks.—Thomas examined two specimens, but identified only the holotype by number; both had been referred tentatively (Thomas, 1897:546) to *Balantiopteryx infusca*, for which name they should not be considered paratypes, when he described that species. The collector was not identified on the label, and the locality was given only as Cobán, Guatemala.

Balantiopteryx plicata Peters

1867.   Monatsb. Kön. preuss. Akad. Wiss. Berlin, p. 476.

ZMB 3361: adult male; in alcohol; Puntarenas, Costa Rica; Osbert Salvin; date of capture not specified. Holotype.

Skin.—Pelage reddish brown; chiropatagium with white border from near fifth digit to foot.

Remarks.—Four specimens (all numbered ZMB 3362) are labeled paratypes (adult male, skin and skull; adult male and two adult females, in alcohol); all four have a white border on the chiropatagium, and the pelage is now reddish brown. The number of specimens examined by Peters was not recorded; however, he reported only a single set of measurements in the original description. The holotype, as well as the four paratypes, were purchased and purportedly came from Puntarenas.

Saccopteryx infusca Thomas

1897.   Ann. Mag. Nat. Hist., ser. 6, 20:546. December.

BMNH (not numbered): adult male; in alcohol; [Río] Cachabi, 500 ft. [in department of Esmeraldas], Ecuador; W. F. H. Rosenberg, no. 30; 5 January 1897. Holotype.

Remarks.—Thomas examined five specimens (two in alcohol plus three skins and skulls) from Cachabi, all collected by Rosenberg on 5 January 1897. Although the two spirit-preserved specimens are without BMNH catalogue numbers and Thomas did not designate the holotype by the collector's number, a label notation on the specimen jar identifies the male (WR 30) as "type." Rosenberg noted on the skin label for paratype 97.11.7.17 that it was taken "from cave in bank of R. Cachabi"; presumably, all came from the same cave. Rosenberg's locality, Cachabi, probably refers to a small river by that name; an elevation of 500 feet would be approximately 6 km. N, 107 km. E Esmeraldas (1°N, 78°40'W). The type locality was spelled "Cachavi" by Thomas.

Diclidurus

Diclidurus albus Wied-Neuwied

1820.   Isis oder Encyclopädische Zeitung von Oken (for 1819), p. 1630.

ZMB 4478: adult of undetermined sex; skin; Rio Pardo [in state of Bahia], Brazil; Freyreiss; date of capture not specified. Holotype.

Skin.—Dorsal and ventral pelage dirty or sooty white, individual hairs on shoulder with basal band of pale buff (on venter, basal half dull reddish, median band of white, terminal band of pale gray); gland on uropatagium as in TCWC 14553.

Skull.—Not located and presumed lost.

Remarks.—Following tradition, we have attributed the name *Diclidurus albus* to Wied-Neuwied in *Isis oder Encyclopädische Zeitung von Oken,* Jena [1817-1848], Jahrgang 1819, p. (actually column number) 1630, although it appears to us that *D. albus* actually is an Oken name. *Isis . . . von Oken* (column 1629) implies that Wied-Neuwied is the author of the name *Diclidurus,* but it is not clear to us that Oken intended to attribute the name *albus* to Wied-Neuwied. In fact, we suspect that the name *albus,* if not the entire account in *Isis . . . von Oken,* was published without the prior knowledge of Wied-Neuwied; no reference to *D. albus* could be found in Wied-Neuwied's (1821) description of *D. freyreisii* although a footnote in *Isis . . . von Oken* indicates that *D. albus* is a synonym of *D. freyreisii.* Wied-Neuwied's first use of the name *albus* appeared in his *Beiträge zur Naturgeschicte von Brasilien . . . ,* 1826; in that work, the name *albus* clearly was attributed to Oken. Wied-Neuwied explained (p. 247) his use of *albus* in place of *freyreisii* and stated that he regretted that Oken applied the name *albus* to the bat obtained by Freyreiss. Schinz (1821:171) also cited Oken (not Wied-Neuwied) for the name *D. albus.* To confuse the matter further, we examined a copy of *Isis . . . von Oken,* Jahrgang 1819, in the British Museum (Natural History) that apparently lacked an account for *Diclidurus albus.*

Diclidurus Freyreisii Wied-Neuwied

1821.   Reise nach Brasilien in den Jahren 1815 bis 1817, 2:76.

ZMB 4478: adult of undetermined sex; skin; Rio Pardo [in the state of Bahia], Brazil; Freyreiss; date of capture not specified. Holotype.

Remarks.—The original Berlin museum catalogue entry (made long after 1826) for ZMB 4478 identified this specimen as a Wied-Neuwied type for *Diclidurus albus,* a senior objective synonym of *Diclidurus freyreisii.* Obviously, ZMB 4478 also is the one that "Herrn Freyreiss in den cocospalmen bei Canavieras am ausflusse des Rio Pardo gefunden" (Wied-Neuwied, 1822-1831, text with pl. 16). [Plates in the British Museum (Natural History) copy of this work were not bound in the order that they appeared in the index to the volume, and probably were not published in the order that they were numbered.]

Diclidurus scutatus Peters

1869.   Monatsb. Kön. preuss. Akad. Wiss. Berlin, p. 400.

MNHN 722: adult of undetermined sex; skin, skull not removed; Amérique du Sud; Barraquin; date of capture not specified. Holotype.

Skin.—Gland on uropatagium as in TCWC 14553 except that sacs are empty.

Remarks.—Peters based this name on a single specimen, which now is sealed in a glass case.

Diclidurus virgo Thomas

1903.   Ann. Mag. Nat. Hist., ser. 7, 11:377. April.

BMNH 98.10.9.3: adult female; skin and skull; Escazú, 3600 ft., Costa Rica; C. F. Underwood; 2 November 1897. Holotype.

Skin.—Torn on dorsum and venter. Dorsum and venter grayish white, venter darker than dorsum; membranes darker than in TCWC 14553; ears yellowish; tragus broad and rounded distally; gland on uropatagium as in TCWC 14553 except empty and flattened.

Skull.—Zygomata broken. Skull like that of TCWC 14553.

Cyttarops

Cyttarops alecto Thomas

1913.   Ann. Mag. Nat. Hist., ser. 8, 11:135. January.

BMNH 12.11.4.5: adult of undetermined sex; skin and skull; Mocajatuba, Pará [=Belem], Pará, Brazil; F. Lima, no. 10, measured 50, 20, 8, 10; 10 May 1912; from Pará museum [Museu Goeldi]. Holotype.

Skin.—Hair on back about 11 millimeters long; dorsal and ventral color slate gray, possibly paler below than above, individual hairs with basal portion paler than distal; membranes and ears dark gray.

Skull.—Postorbital process broken. Rostrum similar to that of *Diclidurus*, not expanded laterally (maxillary toothrow visable from above rostrum); facial ridges weak, no sagittal crest; rostrofacial angle 133°; first upper premolar like that in *Diclidurus*; upper canines with anterolingual and posterolabial cingular styles.

Remarks.—The name "*Cyttarops alecto* gen. et sp. nn." first appears on p. 134, accompanied by a statement on the origin of the holotype, which, taken in the Pará zoological gardens ("em um quintal" is noted on the label) at Mocajatuba (near the city of Pará, now Belem), was obtained from the Museu Goeldi. Thomas examined a second specimen, one presented to the British Museum by F. V. McConnell, whose collector, Cozier, had taken it on the Mazaruni River, British Guiana (now Guyana).

NOCTILIONIDAE

Noctilio

Dirias irex Thomas

1920.   Ann. Mag. Nat. Hist., ser. 9, 6:273. September.

BMNH 20.7.14.29: adult male; skin and skull; Rio Iriri and Rio Xingú, Santa Julia [confluence of rios Iriri and Xingú, Pará, Brazil]; Fräulein E. Snethlage, no. 26, measured 65, 12, 9, 22; 13 February 1914. Holotype.

Skin.—Pale orange-brown above, faint line along midline; pale buff below; chiropatagium dark brown; uropatagium paler brown (possibly faded); ears brown.

Remarks.—The British Museum received this specimen from the Museu Goeldi, Pará. Thomas mentioned a male paratype in his description of *D. irex*, the original number of which was 28 (presumably that of Snethlage), but no other details were given.

Noctilio Albiventer Spix

1823.   Simiarum et vespertilionum brasiliensium species novae . . . , p. 58, pl. 5, figs. 2, 3.

ZSM 17: adult male; skin, skull not removed; Brazil; [apparently Spix]; date of capture not specified. Syntype.

Skin.—Faded.
Skull.—Apparently entire.

ZSM 102/128: juvenile male; skin, skull not removed; [Brazil]; Spix; date of capture not specified. Syntype.

Skin.—Faded. Median line on back whitish.

Remarks.—ZSM 17 was the only adult labeled *Noctilio albiventer* Spix; an old green label is inscribed "*Noctilio albiventer* Spix / (*N. Dorsatus* Neuw. juv.) / Nr. 17 / Brasil." An old label affixed to ZSM 102/128 bears the notation "Original von Spix. / *Noctilio albiventer* (Geoffroy) Spix / *N. leporinus* Linne / juv."

Noctilio rufus Spix

1823.   Simiarum et vespertilionum brasiliensium species novae . . . , p. 57, pl. 35, fig. 1.

ZSM 127: adult male; skin, skull not removed; [apparently Brazil; apparently Spix]; date of capture not specified. Holotype or syntype.

Skin.—Somewhat faded but still rather reddish orange.
Skull.—Apparently entire but part of occiput might be missing.

Noctilio zaparo Cabrera

1907.   Proc. Biol. Soc. Washington, 20:57. 18 April.

MNCN 692: adult female; skin; La Coca Ahuano, Río Napo [in province of Napo], Ecuador; M. Jiménez de la Espada; May 1865. Paratype.

Skin.—Condition poor. Pelage pale reddish above, without median stripe; pale yellow below.
Skull.—Not located and presumed lost.

Remarks.—Although labeled "type," MNCN 692 is one of two paratypes mentioned by Cabrera. The holotype, a male preserved in fluid, number 691, could not be located at the Museo Nacional. All three specimens were stated to have been collected on the Río Napo by Jiménez de la Espada, presumably at the same locality. "Timenez," as the collector's name is spelled in Cabrera's description, is a typographical error.

[*Vespertilio*] *leporinus* Linnaeus

1758.   Systema naturae . . . , ed. 10, 1:32.

BMNH 67.4.12.339: adult male; in alcohol; [Surinam, by restriction (Thomas, 1911:131)]; A. Seba; date of capture not specified. Holotype.

Remarks.—Linnaeus based his description on Seba (1734, vol. 1, p. 89, pl. 55, fig. 1), and the summation of available evidence indicates to us that the British

Museum specimen was once in the possession of Seba and is the holotype for *Vespertilio leporinus* Linnaeus. The label for BMNH 67.4.12.339 has been annotated (possibly by Thomas) "Figd. Seba, pl. LV. fig. 1 therefore Type of species"; this notation was crossed out by Thomas, who added "No. T——said — be ♂ figure is clearly ♀ ." The label is torn, obliterating part of the annotation, but it would seem that Thomas decided that this specimen was not a Linnaean type although he (Thomas, 1892:316) had reported it as such. BMNH 67.4.12.339 obviously is a male whereas the bat figured by Seba appears to be a female, as noted by Davis (1973:871). Notwithstanding, Seba (pp. 89-90) clearly stated in both Latin ("Num. 1. Vespertilio, Cato similis, Americanus; mas.") and French ("No. 1. Chauve-Souris d'Amérique, mâle") that the single specimen in hand was a male. Because the genetalia for male and female noctilionids are similar in character and the art work for the mammals illustrated in Seba's *Locupletissimi verum naturalium thesauri* . . . does not suggest to us any great passion for detail, we conclude that the specimen figured in Seba was a male, and that its feminine appearance is an artifact.

Mormoopidae

Pteronotus

Chilonycteris Davyi fulvus Thomas

1892.   Ann. Mag. Nat. Hist., ser. 6, 10:410. November.

BMNH 93.2.5.24: adult male; skin and skull; Las Peñas, Jalisco, México; Dr. A. C. Buller; 20 November 1891. Holotype.

Skin.—Prepared from fluid-preserved specimen. Pelage on head and shoulder reddish, darker basally; venter paler than dorsum, hair darker basally.
Skull.—Not cleaned.

Remarks.—The generic name *Pteronotus* appears on the label although the name was published originally as *Chilonycteris Davyi fulvus*. Contrary to the information on the skin label, Thomas reported that the holotype was taken on 20 November 1891. The skin label was written at the British Museum, probably when the holotype was prepared as a skin and skull from a spirit-preserved specimen, and we assume that 25 February 1892, as written on the label, is erroneous. Thomas examined five specimens altogether, all taken by Buller at Las Peñas, but identified only the holotype by its British Museum catalogue number.

Chilonycteris gymnonotus Wagner

1843.   Arch. Naturgesch., 9(1):367.

NMW (not numbered): adult female; in alcohol; Cuyabá [Cuiabá, Mato Grosso], Brazil; Johann Natterer, no. 88; date of capture not specified. Holotype.

Skin.—Faded. Pelage now pale reddish brown; chiropatagium inserts on midline of back.
Skull.—Lower incisors strongly trilobed.

Remarks.—Wagner (1843) attributed *Chilonycteris gymnonotus,* and 13 other names first published in his "Diagnosen neuer Arten brasilischer Handfluger," to Johann Natterer; this, together with Wagner's statement (p. 365), "ich von nun an nur noch auf die in sein Tagebuch eingetragen en Notizen beschrankt bin," led Smith (1977) to conclude that Wagner had taken his brief descriptions directly from Natterer's diary and that Natterer alone was responsible both for the names and the conditions that make them available (see International Code of Zoological Nomenclature, Article 50). However, Wagner's (1843) descriptions are not comparable to those found in our photocopy of Natterer's field notes; instead, they appear to have been written independently by Wagner and based largely on specimens received from Joseph Natterer in 1843. Furthermore, Wagner's statement above was made in reference to Johann Natterer having taken an extensive knowledge of the Brazilian fauna to his grave; that is, of this knowledge only Natterer's notes remained. The question of Natterer names is addressed also in our introductory section on authorship.

Chilonycteris MacLeayii Gray

1839.   Ann. Mag. Nat. Hist., 4:5. September.

BMNH 38.6.21.8: adult male; in alcohol; Cuba; W. S. MacLeay; date of capture not specified. Lectotype (Dobson, 1878:450).

Remarks.—The specimen was located but not examined. Technically, this specimen appears to be a lectotype. Gray received altogether two males and one female from MacLeay; none seems to have been designated as a "type" until Dobson (1878:450) did so. *Chilonycteris macleayii* is the type species for *Chilonycteris* Gray (1839:4).

Chilonycteris personata Wagner

1843.   Arch. Naturgesch., 9(1):367.

NMW (not numbered): adult male; in alcohol; São Vicente, Mato Grosso, Brazil; Johann Natterer; date of capture not specified. Syntype.

Skin.—Faded somewhat. Pelage reddish brown above, slightly paler below.

Remarks.—Goodwin's (1946:297) statement that the type locality was probably "St. Vicente," is correct. The only specimen in the Vienna museum with a Natterer label came from São Vicente. According to museum records, there was a second specimen taken at that locality; there are no Natterer specimens of this species from any other locality. Wagner may have kept one of the two specimens, but none was found at Munich.

Chilonycteris psilotis Dobson

1878.   Catalogue of the Chiroptera in the collection of the British Museum, p. 451, pl. 23, fig. 2.

BMNH 50.9.29.3: adult male; in alcohol, skull removed; locality unknown [Tehuantepec, Oaxaca, México, by restriction (Torre, 1955:696)]; Lt. Stricklan; date of capture not specified. Syntype.

Skin.—Condition fair.

Skull.—Much like type of *Pteronotus davyi,* but dorsal depression on rostrum not as deep; I1 bilobed; i1 and i2 trilobed.

Remarks.—The female syntype, recorded by Dobson only as specimen *a,* was not located.

Chilonycteris rubiginosa Wagner

1843.   Arch. Naturgesch., 9(1):367.

ZSM 45: adult male; skin, skull not removed; Caiçara [in state of Mato Grosso], Brazil; Johann Natterer; 19 August 1828. Holotype or syntype.

Skin.—Faded somewhat. Dorsum and venter reddish orange, venter duller than dorsum. *Skull.*—Apparently entire.

Remarks.—The original label, written by Johann Natterer, is inscribed "No. 115 / Caiçara / 19 Aug. 28 / Mas."

Pteronotus Davyi Gray

1838.   Mag. Zool. Bot., London, 2:500.

BMNH 9.1.4.74: age and sex unknown; in alcohol, skull removed; Trinidad, West Indies; collector and date of capture unknown. Holotype or syntype.

Skin.—Condition fair; faded.

Skull.—Skull extracted in 1915. Facial outline concave in side view; rostrum with two longitudinal ridges; I1 bilobed, i1-2 trilobed.

Remarks.—Gray stated that the species "inhabits Trinidad." This specimen once was deposited in the museum of the Royal Army Medical Department at Netley (hence the notation "R.A.M." inscribed on the label) and in the Museum Fort Pitt, Chatham. The name is in honor of John Davy, a well-known physiologist at the time.

Mormoops

Aëllo Cuvieri Leach

1820.   Trans. Linn. Soc., London, 13:71.

BMNH (not numbered): adult of undetermined sex; skin and skull; [presumed to be Jamaica (Smith, 1972)]; W. Bullock (presented to W. Brooke's Museum); date of capture not specified. Holotype or syntype.

Skin.—Dorsal pelage reddish; hairs darker distally; venter paler than dorsum; membranes much faded.

Mormops megalophylla Peters

1864.   Monatsb. Kön. preuss. Akad. Wiss. Berlin, p. 381.

ZMB 2826: adult female; in alcohol, skull removed; Parras [in state of Coahuila], México; Schneider; date of capture not specified. Holotype or syntype.

Skin.—Color reddish, presumably a result of preservation in alcohol.

PHYLLOSTOMATIDAE

PHYLLOSTOMATINAE

Micronycteris

Barticonycteris daviesi Hill

1964.   Mammalia, 28:553.

BMNH 64.767: adult female; in alcohol, skull removed; Forest Reserve, 24 miles from Bartica along Potaro Road, British Guiana [Guyana]; J. N. Davies, no. FM 89; 3 December 1963. Holotype.

Glyphonycteris sylvestris Thomas

1896.   Ann. Mag. Nat. Hist., ser. 6, 18:302. October.

BMNH 96.10.1.2: adult male; skin and skull; Costa Rica [Hacienda Miravalles, 1400-2000 ft., Guanacaste (Goodwin, 1946:302)]; C. F. Underwood; 8 November 1895. Holotype.

Skin.—Dorsal pelage gray, hair on back approximately 9.5 millimeters long, with four color bands (narrow basal band white, gray band, cream band, and dark gray distal band); ventral pelage pale gray, three color bands (basal band white, median band gray, distal band pale gray).

Skull.—Facial-rostral outline slightly concave in side view; rostrum with supraorbital region inflated (width, 5.7); I1 broad, chisel shaped, one-half length of upper canine; I2 crowded, crown not extending beyond cingulum of upper canine; pterygoid pits large, deep, divided medially by well-developed longitudinal partition.

Remarks.—The type locality "Imravalles," as given by Thomas in the original description, is an obvious typographical error. Thomas did not identify the type specimen by number.

Phyllophora megalotis Gray

1842.   Ann. Mag. Nat. Hist., 10:257. December.

BMNH (not numbered): juvenile male; in alcohol, skull removed; Brazil; collector and date of capture unknown. Holotype.

Skin.—Condition moderately poor.

Skull.—As Miller (1907:123-124) described *Micronycteris* (but we cannot see that I1 is divided distally); rostrum with supraorbital region inflated slightly; sagittal ridge weak on anterior face of cranium; basisphenoid pits well defined but smaller and not as deep as in holotype of *Glyphonycteris sylvestris*, divided by median partition; lower incisors with cutting edge trifid rather than bifid as Miller stated for genus.

Remarks.—The specimen is now labeled *Micronycteris megalotis*. Apparently, this is the specimen referred to by Gray (1843:20) as a male preserved in spirits and therefore the holotype of *Phyllophora megalotis*. Part of the problem Gray had in recognizing synonyms for the name no doubt resulted from the immaturity of the holotype of *Phyllophora megalotis*. Of the specimens of

Schizostoma megalotis listed by Dobson (1878:479), the holotype is *g*, not *i* as Dobson indicated.

Phyllostoma elongata Gray

1842. Ann. Mag. Nat. Hist., 10:257. December.

BMNH 42.8.17.8: adult of undetermined sex; skin and partial skull; Brazil; collector and date of capture unknown. Holotype or syntype.

Skin.—Head badly damaged, but ears appear to have been joined on top.
Skull.—Lower incisors trifid.

Remarks.—The notation "Type of *Phyll*[*ostoma*] *elongatum* Gray therefore type of *Phyll*[*ostoma*] *scrobiculatum* Wagner," (author unknown to us) is inscribed on the skull label. However, a specimen in the Vienna museum is labeled type of *Phyllostoma scrobiculatum*. We assume that the note refers to Wagner's (1855:627) statement that "*Phyllostoma elongatum* Gray" is a synonym of *Phyllostoma scrobiculatum* Wagner. Although Dobson (1878:479) did not identify the type of *Phyllostoma elongata*, it was listed as specimen "*f.* ad. sk. Brazil." *Phyllostoma elongata* Gray, 1842, is preoccupied by *Phyllostoma elongatum* É. Geoffroy St. Hilaire, 1810.

Ph[yllostoma]. scrobiculatum Wagner

1855. Die Säugthiere . . . von Schreber, Supplementband, 5:627.

NMW (not numbered): adult male; in alcohol; Brazil; Johann Natterer; date of capture not specified. Holotype or syntype.

Skin.—Faded. Ears joined by obvious fold, approximately 1.5 wide at middle of forehead.
Skull.—Lower incisors weakly trilobed.

Remarks.—A specimen label bears the notation "*Ph. nattereri* Fitzinger." In an attempt to ignore all Fitzinger names, because we believe their recognition could not serve the best interests of mammalian nomenclature, we made no effort to determine the status of *Phyllostoma nattereri*. This specimen also would be the holotype for *P. nattereri* if Fitzinger published the name.

Schizostoma Behnii Peters

1866. Monatsb. Kön. preuss. Akad. Wiss. Berlin, 1865:505.

ZMB 5154: adult female; in alcohol, skull removed; Cuyabá [Cuiabá, Mato Grosso Brazil]; Behn; 2 August 1847. Holotype.

Skin.—Pelage somewhat faded. Individual hairs of dorsum with basal band of white, followed (proximal to distal) by bands of pale reddish brown, cream to pale buff, and reddish brown to gray; venter paler than dorsum, individual hairs with basal band whitish, intermediate band pale buff, terminal bands whitish; ears not joined on top of head.
Skull.—Supraorbital region somewhat inflated; lower incisors weakly trilobed.

Remarks.—Peters examined a single specimen obtained from "Prof. Behn in Kiel."

Schizostoma brachyote Dobson

1879.   Proc. Zool. Soc., London, 1878:880.

MNHN 1876-1074: adult male; in alcohol, skull removed; Cayenne, French Guiana; Mélinon; date of capture not specified. Holotype.

Skin.—Ears not joined across top of head.

Remarks.—Dobson examined a single specimen.

Schizostoma hirsutum Peters

1869.   Monatsb. Kön. preuss. Akad. Wiss. Berlin, p. 396.

MNHN (not numbered): adult male; in alcohol, skull removed; exact locality unknown [Pozo Azul, Guanacaste, Costa Rica, by restriction (Goodwin, 1946: 302)]; collector and date of capture unknown. Holotype.

Skin.—Pelage dark brown; ears connected, or almost so by band across top of head, notch in center of band.

Skull.—Zygomata broken, otherwise condition good.

Remarks.—This specimen was listed as type 189 in Rode's (1941) catalogue. Peters examined a single specimen.

Lonchorhina

Lonchorhina aurita Tomes

1863.   Proc. Zool. Soc., London, p. 83, pl. 12.

BMNH 9.1.4.67: subadult of undetermined sex; in alcohol, skull removed; W. Indies, Trinidad?; date of capture not specified. Holotype.

Remarks.—Type locality is purported to be Trinidad (see Thomas, 1893; Hall and Kelson, 1959:104). The notation "R.A.M." on the label identifies this specimen as once having been housed in the museum of the Royal Army Medical Department at Netley. According to Tomes, this specimen was part of a collection of bats in the museum at Fort Pitt, Chatham (it would seem most probable that the museum at Ft. Pitt, part of a major British military installation, obtained the specimen from the R.A.M.). Tomes stated that the specimen was taken from a bottle containing "several West Indian species, in which the *Mormops blainvillii* and the *Chilonycteris gymnonota* [= *Pternontus davyi*] of Wagner were conspicuous." For this reason, (and probably because *Mormoops blainvillii* was known to occur only in the West Indies), Tomes suggested that the locality for his specimen of *L. aurita* was probably the same as that of "the *Mormops* and *Pteronotus*" (that is, the West Indies). Although *Pteronotus davyi* occurs on Trinidad and in the Lesser Antilles as far north as the island of Dominica, *Mormoops blainvillii* presently is restricted to the Greater Antilles (Smith, 1972:109). For what it is worth, *Pteronotus davyi* Gray, 1838, also from Trinidad, was once housed in the museums at Netley and Ft. Pitt.

The epiphyses of the metacarpals and phalanges are almost closed; the specimen is damaged in such a way that the sex cannot be determined (J. E. Hill, personal communication).

Tonatia

Lophostoma brasiliense Peters

1866.   Monatsb. Kön. preuss. Akad. Wiss. Berlin, p. 674.

BMNH 49.11.7.14: adult female; skin and skull; Bahia, Brazil; [purchased from] Brandt; date of capture not specified. Holotype.

Skin.—Dorsal pelage gray, hair with proximal one-fourth white, distal three-fourths gray; ventral pelage paler than dorsal, hair with basal band white, distal band gray; membranes, ears, and nose leaf gray; ears shaped like those of Tonatia childreni.

Skull.—Rostrum and lower jaw only. Lower incisors weakly bifid; first lower premolar with posterior portion overlapping anterior part of second lower premolar; second lower premolar with cutting edge slightly above anterior border of cingulum of third lower premolar; third lower premolar with point of cusp in center of tooth.

Remarks.—Peters' name, Lophostoma brasiliense, is a recombination and description of Tylostoma brasiliense Gray, a nomen nudum. Brandt was a dealer in natural history objects.

Phyllostoma amblyotis Wagner

1843.   Arch. Naturgesch., 9(1):365.

NMW (not numbered): adult female; in alcohol; Mato Grosso, Brazil; Johann Natterer; date of capture not specified. Syntype.

Skin.—Faded. Dorsum pale brown, with slight reddish cast, individual hairs white at base, tips frosted as in TCWC 11702 and 11705; venter white to very pale brown (except for fading, color as in TCWC 11705, but the gray color in 11705 is very pale brown in the type specimen of P. amblyotis); ears, tragus, and nose leaf as in TCWC 11705.

Skull.—Like that of TCWC 11705 except that lower incisors faintly bilobed, and second lower premolar larger, the crown extending slightly above posterior margin of crown of first lower premolar.

Remarks.— Wagner examined two specimens.

Phyllostoma Childreni Gray

1838.   Mag. Zool. Bot., London, 2:488.

BMNH 8a: adult male; skin and skull; South America; J. G. Children; date of capture not specified. Holotype.

Skin.—Possibly prepared from spirit-preserved specimen. Dorsal pelage pale brown with narrow basal band white, median band gray, distal band pale brown; ventral pelage gray brown with cream frosting; ears not joined medially, measured (dry) 18, much shorter and narrower than in Tonatia laephotis and TCWC 11705.

Skull.—Rostrum perhaps two-thirds length of braincase, relatively short and broad in appearance; interorbital region much broader than in Tonatia laephotis; lower incisors weakly bifid, second lower premolar overlapped by cingulum of first and third lower premolars; third lower premolar with point of main cusp in center of crown or possibly slightly posterior to center of crown; lower premolars similar to those of TCWC 11705, but the rostrum of BMNH 8a is much broader than that of TCWC 11705; sagittal crest weakly developed.

Remarks.—Gray examined a single specimen.

Tonatia laephotis Thomas

1910.   Ann. Mag. Nat. Hist., ser. 8, 6:184. August.

BMNH 10.5.4.5: adult of undetermined sex; skin and skull; Río Supinaam [a tributary of the lower Río Essequibo], West Demerara, British Guiana, [Guyana]; Cozier, prepared by R. V. McConnell; date of capture not specified. Holotype.

Skin.—Dorsal pelage dark gray with cream (or whitish) frosting (narrow basal band white); ventral pelage paler than dorsal (hairs with basal band white, median band gray, distal band white or cream); membranes, ears, and nose leaf gray; ears not joined medially.

Skull.—Rostrum about three-fourths length of braincase; relatively narrow and elongate; interorbital region narrow; lower incisors bifid. Compared with TCWC 11705 and 11702, holotype of *Tonatia laephotis* has skull larger; lower incisors bifid instead of entire; second lower premolar extending above cingulum of third lower premolar and not overlapped dorsally by first lower premolar; third lower premolar with primary cusp directly over anterior root instead of in center of crown (and therefore above space between anterior and posterior roots).

Remarks.—As originally printed, the diphthong *ae* in *laephotis* was set as a ligature in a typeface that caused it to look much like the diphthong *oe*. For that reason, the name sometimes has been misspelled *loephotis* (for example, Goodwin, 1942). Thomas was silent on the etymology of the specific epithet, but he certainly must have taken it from the Greek λαιόσ (properly Latinised as *laeus* and meaning left-handed, awkward, or clumsy—Laius, the father of Oedipus, was left-handed and had an awkward or clumsy gait) plus φῶσ (Latinised as *phos* and meaning light, as in sunlight). In combination, the two words λαιόσ φῶσ would indicate clumsy movement or flight in daylight. Thomas examined a single specimen.

Vampyrus Bidens Spix

1823.   Simiarum et vespertilionum brasiliensium species novae . . . , p. 64, pl. 36, fig. 5.

ZSM (not numbered): adult of undetermined sex; skin and skull; Rio São Francisco, Bahia, Brazil; [apparently Spix]; date of capture not specified. Holotype or syntype.

Skin.—Dorsal pelage dull brown, tips of hairs paler; ventral pelage paler than dorsal; ears relatively small, not joined; forearms with proximal half densely furred.

Skull.—Frontal and rostral slope essentially continuous; il weakly bilobed, outer lobe smaller; cingula of first and third lower premolars extend above crown of second lower premolar.

Remarks.—Although there is no other specimen in the Zoologisches Staats-Sammlung München labeled *Vampyrus bidens* (or labeled in any other way that would indicate that it was one on which this name was based), the skin and skull are unnumbered and there is no way to ascertain that they belong together.

Mimon

Anthorhina peruana Thomas

1923.   Ann. Mag. Nat. Hist., ser. 9, 2:693. December.

BMNH 23.10.16.12: juvenile female; skin and skull; Río Pachitia, 1500 ft. [in department of Huánuco], Perú; Latham Rutter, no. 80; date of capture not specified. Holotype.

Skin.—Dorsum blackish, with narrow white stripe, individual hairs with narrow basal band white or very pale gray; venter gray, with white frosting, narrow basal band white; membranes blackish; ears blackish distally, conch and tragus yellowish; nose leaf dark distally, paler basally.

Skull.—Lower incisors bifid, crown longer than wide (in anterior view).

Remarks.—Thomas erroneously reported this specimen to be an adult female. It probably was taken in 1923 (possibly around October) inasmuch as Rutter is known to have been collecting in Perú at that time.

Anthorhina picata Thomas

1903.   Ann. Mag. Nat. Hist., ser. 7, 12:457. October.

BMNH 3.9.5.26: adult male; skin and skull; Lamarão, 300 m., Bahia, Brazil; Alphonse Robert, no. 1410, measured 65, 22, 8, 27; 24 March 1903. Holotype.

Skin.—Dorsal pelage dark gray (narrow basal band white), longitudinal white stripe from head to base of tail, ears with white hair at posterior base; ventral pelage cream or very pale buff (narrow basal band white, median band gray, distal band pale buff or cream); nose leaf, ear conch, and tragus yellowish basally, gray distally.

Skull.—Lower incisors trifid, crowns about as broad as high in anterior view, converging distally; cingulum defined on anterior surface of tooth.

Remarks.—The date of capture on the specimen label appeared to be 24 March; however, Thomas reported the date as 24 May.

Phyllostoma crenulatum É. Geoffroy St.-Hilaire

1810.   Ann. Mus. Nat. Hist. Nat. Paris, 15:183, pl. 6.

MNHN (not numbered): adult male; in alcohol, skull removed; [Bahia, Brazil, by restriction (Cabrera, 1958:66)]; collector and date of capture unknown. Holotype or syntype.

Skin.—Condition fair, faded. As figured by É. Geoffroy St.-Hilaire.
Skull.—Condition poor, deformed from preservation in fluid.

Remarks.—The locality for this specimen was unknown to É. Geoffroy St.-Hilaire, who stated only that it was "Amérique vraisemblablement."

Phyllostomus

Phyllostoma discolor Wagner

1843.   Arch. Naturgesch., 9(1):366.

ZSM 133: adult male; skin, skull not removed; Cuyabá [Cuiabá, Mato Grosso], Brazil; Johann Natterer, no. 96; September 1824. Holotype or syntype.

Skin.—Faded. Pelage dull reddish brown on most of back.

Skull.—Rostrum and mandible only. Incisors and canines as in TCWC 10734, 11725, 11744.

Remarks.—The specimen bears an original label inscribed "no. 96 / Cuyaba / Septb 24 / Mas."

Phyllostoma elongatum É. Geoffroy St.-Hilaire

1810.   Ann. Mus. Nat. Hist. Nat. Paris, 15:182, pl. 5.

MNHN A.2: adult female; in alcohol, skull removed; [Rio Branco, Mato Grosso, Brazil, by restriction (Cabrera, 1958:67)]; collector and date of capture unknown. Holotype or syntype.

Skin.—Faded.

Remarks.—The locality for this specimen was unknown to Geoffroy St.-Hilaire, who stated (p. 185) that it was from "Amérique, selon toute apparence." The type of *Phyllostoma elongatum* is entry 190 in Rode (1941).

Phyllostoma latifolium Thomas

1901.   Ann. Mag. Nat. Hist., ser. 7, 8:142. August.

BMNH 1.6.4.43: adult male; skin and skull; Kanuku Mts., 600 ft. [in Guyana]; J. J. Quelch (prepared by F. V. McConnell), no. 199, measured 76, 15, 12.5, 25, expanse 442; 4 December 1900. Holotype.

Skin.—Dorsal pelage brown, narrow basal band white or pale gray; ears and nose leaf yellow brown; membranes blackish brown; venter paler than dorsum, tips of hair paler than proximal portion.

Skull.—Rostrum short (approximately 10.5), two-thirds length of braincase; postorbital process present and distinct; sagittal crest present; I1 and i1-2 weakly bilobed.

Remarks.—In addition to the holotype, Thomas examined five skins and two spirit-preserved specimens. These were collected at altitudes of 1000 feet on 6 November and at 600 feet on 4 December 1900. With the exception of external measurements provided for one fluid-preserved male, no other details for the paratypes were given, and they were not identified by number.

Phylloderma

G[*uandira*]. *cayanensis* Gray

1866.   Proc. Zool. Soc., London, p. 114.

BMNH 42.10.25.2: adult male; skin and partial skull; Cayenne [in French Guiana]; collector and date of capture unknown. Holotype or syntype.

Skin.—Dorsal pelage brownish, with basal band white; ventral pelage with basal band brown, distal band white; membranes, ears, and nose leaf brownish.

Skull.—Incomplete. Teeth as in TCWC 16412 from Chiapas, but smaller.

Ph[ylloderma]. stenops Peters

1865.   Monatsb. Kön. preuss. Akad. Wiss. Berlin, p. 513.

RNH 16843: adult, probably male; mounted skin, skull removed; Cayenne [in French Guiana]; collector and date of capture unknown. Holotype.

Skin.—Faded. Pelage banded as in TCWC 16412 (*Phylloderma* from Chiapas); nose leaf shorter and narrower than in TCWC 16412 (measured from the back, they are the same, but held next to each other 16412 is noticeably longer); ears shorter than in TCWC 16412.

Skull.—Similar in form and proportions to TCWC 16412 but much smaller; upper premolars less crowded (first and second upper premolars barely overlapping; lower premolars less crowded; teeth smaller than in 16412.

Remarks.—The wooden base on which the skin was mounted is inscribed male, and the recent notation "ad. male" accompanies the specimen. However, the label for the wooden base indicates that the holotype is a female. The sex cannot be determined with certainty, and what appears to be part of the penis may only be a piece of skin.

Trachops

Trachops cirrhosus ehrhardti Felten

1956.   Senck. Biol., 37:369.

SMF 11716: adult female; in alcohol, skull removed; Joinville, Santa Catarina, Brazil; W. Ehrhardt; 1908. Holotype.

Skull.—First and second lower incisors trilobed.

Remarks.—There are two paratypes (11717 female and 11718 male), both of which are adults.

Chrotopterus

Chrotopterus auritus australis Thomas

1905.   Ann. Mag. Nat. Hist., ser. 7, 16:308. September.

BMNH 1.3.11.1: adult male; skin and skull; Concepción [in department of Concepción], 300 m., Paraguay; J. Insley, no. 6, measured 120, ——, 22, 44; 6 May 1900. Holotype.

Remarks.—The notation "captured in bedroom during night" appears on the skin label. Thomas stated that several specimens also were "obtained in Paraguay by Mr. W. Foster," but it is not known whether he used them in describing *Chrotopterus auritus australis.*

Chrotopterus auritus guianae Thomas

1905.   Ann. Mag. Nat. Hist., ser. 7, 16:308. September.

BMNH 4.5.7.20: adult male; skin and skull; La Vuelta, lower Orinoco, Venezuela; S. M. Klages, no. 364; 24 April 1903. Holotype.

Vampyrus auritus Peters

1856.   Monatsb. Kön. preuss. Akad. Wiss. Berlin, p. 415.

ZMB 3755: adult male; in alcohol; Santa Catarina, Brazil; H. Burmeister; date of capture not specified. Holotype or syntype.

Skin.—Pelage now reddish in color, but back with grayish cast and hairs tipped with white.

Remarks.—Peters' apparent reference to *Vampyrus auritus* as the "verwandten Art aus Mexico" naturally led subsequent authors to assume that the name was based on one or more specimens from México; all existing label information for ZMB 3755, taken by Burmeister in Santa Catarina, indicates that it is a type specimen for *Vampyrus auritus.* There is, however, a spirit-preserved specimen (male) in the Berlin museum, the label for which indicates only that it came from México.

Glossophaginae

Glossophaga

Glossophaga amplexicaudata Spix

1823.   Simiarum et vespertilionum brasiliensium species novae . . . , p. 66, pl. 36, fig. 4.

ZSM 191: two adults and three juveniles of undetermined sex; in alcohol; locality, collector, and date of capture unknown. Syntypes.

Remarks.—Although in a sealed bottle, which was not opened, we are reasonably certain that the two adults are *Glossophaga soricina* inasmuch as the mouths are open and reveal a procumbent I1 (broadened distally) and lower incisors that form a continuous, essentially horizontal (in anterior view) row between the canines. An old white label is inscribed "4 stuck / *Glossophaga amplexicaudata* (Spix) / A.M. 83-86 / *typus* / Spix"; an old green label, "*Glossophaga amplexicaudata* Spix."

In addition to the above, there is an unnumbered specimen (adult of undetermined sex, possibly a female; mummified, skull in place; Brazil; collector and date of capture not specified) identified with an old green label as "*Glossophaga amplexicaudata* Geoffroy / Brasil." This specimen long has been in the Zoologisches Staats-Sammlung München collection, but there is no indication that it has any type status.

The name *Glossophaga amplexicaudata* Spix (a species of *Glossophaga*) seemingly was confused with *Glossophaga amplexicauda* É. Geoffroy St.-Hilaire (a species of *Carollia*), which most authors routinely misspelled *amplexicaudata*.

Monophyllus Leachii Gray

1844.   The zoology of the voyage of H.M.S. Sulphur . . . , 1(Mammalia):18, April.

BMNH 42.8.17.17: adult of undetermined sex; skin and skull; Realejo [in department of Chinandega], Nicaragua; J. Gould; date of capture not specified. Holotype.

Skin.—Condition poor. Dorsal pelage brown, hairs with basal band (two-thirds of hair shaft) pale brown, distal band dark brown with pale brown frosting; venter pale yellowish brown, hairs darker distally, tipped white or cream; membranes, ears, and nose leaf pale brown (perhaps faded).

Skull.—Damaged, occiput and most of basal portion of cranium missing. Rostrum approximately one-half length of braincase; upper incisors appear less procumbent than is usual for *Glossophaga soricina*; I1 expanded distally, bilobed, longer than I2; I1 and I2 in contact; I2 separated from canine by space that is about one-half cingular diameter of I2; upper canines with minute anteromedial cingular style and posterior cingular style on small posterior cingular shelf; upper premolars two in number and double rooted, central cusp slightly anterior to center of tooth, and anterior and posterior cingular styles small; second upper premolar larger than first; upper molars with well-developed central depression (between protocone, paracone, and metacone); M1 largest of molars, protocone, paracone, and metacone well developed, parastyle much reduced, metastyle rather well developed (ectolophs much compressed laterally); M2 with prominent protocone, paracone, and metacone, parastyle well developed, metastyle much reduced; M3 smallest of molar series, protocone, paracone, and metacone rather well developed, parastyle present and well developed, metastyle absent, smallest of the three molars; upper toothrow with small spaces between canine and first premolar and between first and second premolar, second and third premolars in contact; lower incisors small, rather peglike, slightly expanded distally, not in contact; i2 separated from canine by small space; lower canine with small posterior shelf; first and second lower premolars with one main cusp situated on anterior part of tooth, small posterior cingular style; third lower premolar with anterior and posterior cingular style; first and third lower premolars approximately equal in size; second lower premolar slightly smaller than first and third; no space between teeth; m1 largest of molar series, paraconid smaller than protoconid and metaconid and arranged along midline of tooth, hypoconid well developed and larger than entoconid; m2 smaller than m1, with relatively large paraconid on lingual side; m3 like m2 except that hypoconid and entoconid united (or apparently so).

Phyllostoma soricinum É. Geoffroy St.-Hilaire

1810.   Ann. Mus. Nat. Hist. Nat. Paris, 15:179, pl. 11.

MNHN A. 278: juvenile male; in alcohol, skull removed; Surinam; collector and date of capture unknown. Holotype.

Skin.—Faded; bones limber.

Skull.—Recently removed; deformed due to long preservation in alcohol. Permanent teeth incompletely erupted, but I1 obviously longer than I2 and procumbent; lower incisors adjacent (not in pairs) and evenly distributed between canines.

Remarks.—Pallas (1766) is the author of the name *Vespertilio soricinus*. *Phyllostoma soricinum*, although based on a specimen from Surinam and labeled type, appears simply to be a redescription and reassignment of *V. soricinus* Pallas, for which É. Geoffroy St.-Hilaire (1810:179; 1818a:418) indicated that it was a synonym. MNHN A. 278 was examined but not measured.

Lonchophylla

Lonchophylla mordax Thomas

1903. Ann. Mag. Nat. Hist., ser. 7, 12:459. October.

BMNH 3.9.5.34: adult male; skin and skull; Lamarão, NW of Bahia [=Salvador], 300 m., Bahia, Brazil; Alphonse Robert, no. 1552, measured 55, 8, 10, 13, expanse 260, ("expanse 265" according to Thomas); 3 July 1903. Holotype.

Skin.—Hair on back bicolored, basal two-thirds pale reddish gray, distal one-third reddish brown; venter pale gray to pale reddish, with hairs tipped with pale gray.

Skull.—Braincase evenly rounded, longer than rostrum; second (and last) upper premolar with distinct internal, medial cusp (tooth obviously three rooted); M1, M2, and M3 with parastyle and metastyle well developed; M3 with parastyle well developed, metastyle absent; upper molars three rooted, posterointernal root perhaps weak; lower incisors gently convex in dorsal aspect; i1 and i2 in contact; diastema between i2 and canine; first lower premolar without anterior style, posterior style present; second and third lower premolars with anterior and posterior styles present (all three premolars two rooted); m1 and m2 about equal in size, larger than m3; lower molars with paraconid weakly developed, otherwise tuberculo-sectorial; lower molars apparently two rooted.

Remarks.—There are seven paratypes in the British Museum.

Lionycteris

Lionycteris spurrelli Thomas

1913. Ann. Mag. Nat. Hist., ser. 8, 12:271. September.

BMNH 13.8.10.1: juvenile male; skin and skull; Condota, 300 ft., Chocó, Colombia; Dr. H. G. F. Spurrell, no. 314, measured 49, 7, 9, 13; 10 May 1913. Holotype.

Skin.—Hair on back dark gray basally, changing to dark reddish brown distally; hair on venter medium gray basally, pale rusty to pale gray distally; membranes dark gray; ears and nose leaf pale gray (probably faded).

Skull.—Rostrum, one-half to two-thirds length of braincase, relatively broad and deep; braincase evenly expanded and rounded dorsally; palate extends caudad about one-half the distance from M3 to glenoid fossa, posterior border without notches, small projection in center of interpterygoid space; pterygoid not much deflected laterally and not in contact with tympanum; I1 expanded distally; milk I2 present on right side and recurved, permanent I2 can be seen in alveolus; permanent I2 not yet emerged on left side; upper canine small, distinct, posterior cingular style present; first upper premolar separated from canine by a space equal to diameter of premolar, with large primary cusp and small distinct posterolingual style triangular in cross section; M1 with protocone, metacone, paracone, parastyle well developed, mestastyle reduced; M2 like M1; M3 lacks metastyle although cingular shelf present on labial side of paracone, otherwise like M2; first two molars approximately equal in size, third a little smaller; lower incisors trilobed, more or less filling space between canines, but i2 not in contact with canine; lower canine with small antero and posterolingual cingular styles, canine separated by narrow space from first lower premolar; lower premolars two rooted, with central cusp relatively long and large, anterior and posterior styles relatively small but distinct; second premolar slightly larger than first, third slightly larger than second; lower premolars not compressed laterally, relatively broad cingular style on lingual side of tooth; lower molars three rooted, five cusps on m1, paraconid reduced but distinct, situated on midline of m1, paraconid on lingual side of midline on m2-3; talonid well developed and depression well defined; m1 is largest, m2 and m3 progressively smaller.

Remarks.—No differences were found between the holotype and TCWC 11900. The type locality was printed as "Condoto, Choco, Colombia. Alt. 300'." Thomas based his description on a single specimen.

Anoura

Anoura Geoffroyi Gray

1838. Mag. Zool. Bot., London, 2:490.

BMNH 11a: adult of undetermined sex; skin and skull; Brazil; collector and date of capture unknown. Holotype.

Skin.—Condition poor. Hair on back pale yellow brown on basal two-thirds, dark reddish brown on distal one-third or one-fourth; membranes and ears pale yellow brown.

Skull.—Rostrum approximately three-fourths length of braincase; braincase evenly expanded and gently rounded dorsally; I1 peglike, not in contact with I2; I2 two-times size of I1, extends beyond I1, separated from canine by a distance equal to the greatest diameter of I2; canine with small, posterior cingular style; first upper premolar small, about one-half size of second, two rooted, with primary cusp and posterior style; second upper premolar with anterior and posterior styles, primary cusp slightly anterior to center of tooth; third upper premolar two rooted, larger and less compressed than second, anterior and posterior styles present, primary cusp approximately in center of tooth; upper molars more or less triangular in occlusal view; M1 and M2 more or less equal in size, M3 smaller, approximately three-fourths or four-fifths size of M2; all three molars with protocone, paracone, metacone, parastyle, and metastyle well developed and central depression (between the three primary cones) well defined; M1 and M2 with cingular style posterior to metastyle; M3 without third style; palate with notch posterior to anterior zygomatic process of maxillary, small projection into interpterygoid space, not extending to glenoid fossa; pterygoid depressed laterally, not in contact with tympanum; lower incisors absent; lower canine with weak posterointernal cingular shelf; first lower premolar with primary cusp (comprising anterior two-thirds of tooth) and posterior style; second lower premolar slightly larger than first, with primary cusp (more or less in center of tooth) plus anterior and posterior styles; third lower premolar like second but slightly larger; lower molars with five cusps, but paraconid much reduced and situated along midline of tooth; talonid well developed, depression well defined; m1 largest, m2 slightly smaller than first and slightly larger than third.

Remarks.—The specimen is an adult although Dobson (1878) stated that it was an "immature adult." Sanborn (1933:26) reported the type locality to be Rio de Janeiro?" probably because *Glossophaga ecaudata* É. Geoffroy St.-Hilaire (a name which Gray supposed to be synonymous with his *Anoura geoffroyi*) was based on one or more specimens taken at that place by Delalande fils.

Glossonycteris lasiopyga Peters

1868. Monatsb. Kön. preuss. Akad. Wiss. Berlin, p. 365, fig. 2.

ZMB 3564: adult of undetermined sex; skin; Cuernavaca, "12 meilen von Mexico," [in state of Morelos], México; Boucard; August 1866. Syntype.

Skin.—Faded somewhat.
Skull.—Not found and presumed lost.

ZMB 3565: adult male; skin, skull not removed; Cuernavaca, "12 meilen von Mexico" [in state of Morelos], México; Boucard; August 1866. Syntype.

Skin.—Faded somewhat.

Remarks.—Although Peters referred to a single dried specimen (obtained from Saussure), for which he gave external measurements, both ZMB 3564 and 3565 are labeled "typus." Because the skull was figured, one might assume that ZMB 3564 was the one to which the description refers, but the measurements given by Peters do not appear to be those of this specimen. We consider both to be syntypes of *Glossonycteris lasiopyga*. We found no other Mexican specimens of *Anoura* in the Berlin museum.

Glossonycteris lasiopyga is the type species for *Glossonycteris* Peters.

Glossophaga caudifer É. Geoffroy St.-Hilaire

1818.   Mem. Mus. Hist. Nat. Paris, 4:418, pl. 17.

MNHN 937: adult female; skin; vicinity of Rio de Janeiro, Brazil; Delalande fils and Auguste de St.-Hilaire; date of capture not specified. Holotype or syntype.

Skin.—Faded. Uropatagium and legs less hairy than on TCWC 11886, which has a dense fringe of hair on margin of uropatagium.

Skull.—Not located and presumed lost. No reference was made to a skull in Geoffroy's description.

Remarks.—The Paris museum's A-series catalogue of *circa* 1868 indicates that this specimen came from Auguste de St.-Hilaire, 1822. The date 1822 is probably the date of registration and not that of collection or receipt by the Paris museum. Notwithstanding Geoffroy's (1818a:418) statement, "espece nouvelle, du voyage de M. Delalande fils," the type (or syntypes) for *G. caudifer* was received from "MM. Delalande fils et Auguste de St.-Hilaire" (É. Geoffroy St.-Hilaire, 1818a:411).

Lonchoglossa wiedi aequatoris Lönnberg

1921.   Ark. Zool., 14(4):65. 7 June.

NR 6: adult male; skin and skull; "Ilambo, Gualea, 5000 ft.," Ecuador; collector not identified; 20 April 1913. Holotype.

Skin.—Dorsal pelage dark gray (individual hairs medium gray basally becoming dark gray distally); membranes, ears, and nose leaf blackish; uropatagium narrow, hair on dorsal surface, fringe of hair on margin (density of hair on margin as in TCWC 11883 and 11886); ears and tragus similar to those of TCWC 11883 and 11886.

Skull.—Rostrum approximately three-fourths length of braincase; frontal region rises rather abruptly from rostrum; skull and teeth similar to those of TCWC 11883, except that facial angle more acute and rostrum broader in holotype.

Remarks.—There is one paratype (labeled cotype), NR 8, an adult male taken at "Ilambo, Gualea, 5000 ft.," on 24 April 1913.

Scleronycteris

Scleronycteris ega Thomas

1912.   Ann. Mag. Nat. Hist., ser. 8, 10:405. October.

BMNH 7.1.1.671: adult female; skin and skull; Ega, Amazonas, Brazil; H. O. Bates, no. 171; date of capture not specified. Holotype.

Skin.—Hair on back approximately 8 millimeters in length, pale brown on basal two-thirds and dark brown on distal one-third or one-fourth; hair on venter pale yellow brown on basal three-fourths and reddish brown on distal one-fourth; membranes, ears, and nose leaf dark brown; nose leaf broad and short, 3.8 (measured from upper lip to top of leaf—upper point of leaf seems to have been broken off).

Skull.—Damaged, occiput and basal portion missing; specimen originally prepared without removing skull. Rostrum shorter than braincase; braincase evenly expanded, rises gently from rostrum in facial region; upper incisors small, otherwise as in Thomas' description; upper canines with moderate posterior basal ledge; upper premolars two rooted and moderately compressed laterally, anterior style greatly reduced, posterior style more prominent; molars with protocone and metacone well developed, but paracone much reduced and appears to be an almost stylelike projection from parastyle; parastyle and metastyle well developed; molars with central depression as in *Glossophaga* and *Lonchophylla*; upper canine and first premolar separated by a space about as long as linear dimension of first premolar; premolars and molars separated by a narrow space; lower incisors absent (strong anterior, downward projection of mandibular symphysis); lower canines with strong posterior basal ledge; lower premolars triconate, anterior and posterior styles approximately equal in development; first lower premolar with central cone only slightly stronger than anterior and posterior styles, second and third lower premolars with much stronger central cusps; lower molars five cusped, only moderately compressed laterally, talonid well developed, paraconid weak and shifted labially.

Remarks.—The holotype for *Scleronycteris ega* is from the Tomes collection, number 212a.

Lichonycteris

Lichonycteris obscura Thomas

1895.   Ann. Mag. Nat. Hist., ser. 6, 16:56. July.

BMNH 95.4.27.1: adult female; in alcohol, skull removed; Managua [in department of Managua], Nicaragua; Dr. E. Rothschuh; date of capture not specified. Holotype.

Skin.—Condition reasonably good; some fading. Nose leaf relatively short and broad.

Skull.—Rostrum short, rather narrow, approximately two-thirds length of braincase; braincase rather evenly expanded and rounded; posterior margin of palate entire (incised in TCWC 9831); upper incisors small, I1 bilobed and compressed, I2 round in cross-section and longer than first; upper premolars two rooted; first upper premolar with anterior style weak, posterior style moderately strong; second upper premolar with anterior style reduced (but stronger than in first premolar), posterior style forming weak shelf; upper molars apparently three rooted; M1 moderately compressed laterally (but depression in center of tooth well developed, as in *Glossophaga*), with protocone, paracone, and metacone moderately well developed, parastyle and metastyle similar to paracone and metacone in development; M2 like M1 except not compressed and shorter than first, volume about equal in the two; lower incisors absent; lower premolars triconate (central cusp best developed of the three, anterior style less developed than posterior style), separated from each other by a space approximately equal to one-third or one-half anteroposterior length of tooth, space separating third premolar from first molar narrow; lower molars two rooted, with three cusps, paraconid less developed than protoconid and metaconid, paraconid shifted labially from original position, talonid well developed; m2 approximately two-thirds size of m1.

Remarks.—Thomas based his description on a single specimen, and the name of this species first appears on page 55 as "*Lichonycteris obscurus*, gen. et sp. n."

Hylonycteris

Hylonycteris Underwoodi Thomas

1903.   Ann. Mag. Nat. Hist., ser. 7, 11:287. March.

BMNH 3.2.1.5: adult of undetermined sex; skin and skull; Rancho Redondo, San José, Costa Rica; C. F. Underwood; 30 June 1899. Holotype.

Skin.—Hair on back 8 millimeters in length dense and soft, medium dark gray basally, pale gray to very pale buff on midportion, dark brown on distal one-fourth or one-fifth; distal part of hair paler brown (with reddish cast) on sides; hair on venter gray on basal half, rusty brown distally, some with pale tip.

Skull.—Rostrum three-fourths to four-fifths length of braincase, relatively broad but shallow; face slopes gently; braincase more or less evenly rounded; posterior border of palate V-shaped, without lateral palatal notches, even with anterior margin of glenoid fossa; pterygoid inflated, not in contact with tympanum, deflected laterally but not extremely so; I1 faintly bilobed; I2 cone shaped, the two upper incisors approximately equal in size (I1 may be slightly smaller); upper canine without posterior cingular style but cingulum present (not actually in form of a shelf); upper premolars two rooted, two in number: first upper premolar about two-thirds or three-fourths size of second, with primary cusp slightly anterior to center, anterior and posterior styles weak; second upper premolar laterally compressed, with central cusp slightly anterior to center of tooth, anterior and posterior styles weak, more shelflike than stylelike; upper molars three rooted (missing from right side, worn on left side); M1 with protoconid, metaconid, and apparent paraconid, weakly defined central depression, parastyle absent (may have been broken off), metastyle present and directed caudad; M2 like M1, approximately same size; M3 similar to M2 but metastyle and parastyle weakly developed, smaller than M2; lower incisors absent; lower canine with weak posterior cingular shelf; lower premolars three in number, all approximately equal in size and almost in contact (canine in contact with first premolar), primary cusp relatively small, slightly anterior to center of tooth, anterior and posterior styles somewhat shelflike (possibly due to wear); lower molars laterally compressed; m1 with five cusps but paraconid weak and on labial side of midline, talonid moderately developed with moderately defined depression; m2 like m1, but approximately three-fourths size of m1, paraconid situated on midline of tooth or slightly labial to midline; m3 slightly more than one-half size of m1, paraconid as in m2, talonid shelflike (hypoconid and entoconid very weak and talonid basin ill defined).

Remarks.—Compared with holotype, TCWC 7303 has a narrower, shallower, rostrum, which is shorter in relation to braincase; also, the facial angle is less acute, giving braincase a more evenly rounded appearance. The two specimens appear to represent closely related populations, but TCWC 7303 has a smaller and more delicate skull than does BMNH 3.2.1.5. The two skins are similar.

The name *Hylonycteris underwoodi* first appears on page 286 as the type species for *Hylonycteris*, but the description of this species is on page 287. Thomas examined two paratypes from Tarbaco, San José, Costa Rica, also taken by C. F. Underwood.

Platalina

Platalina genovensium Thomas

1928.   Ann. Mag. Nat. Hist., ser. 10, 1:121. January.

BMNH 27.11.19.38: adult male; skin and skull; near Lima, Perú; Nicolo Esposta, measured 72, ——, 9, 13, "Av. 48, P.p. 9.5"; 23 April 1909. Holotype.

Skin.—Prepared from a fluid-preserved specimen. Pelage pale buff, hairs becoming white basally; membranes reddish brown; tail present; calcar relatively well developed; uropatagium relatively wide.

Skull.—Broken across rostrum (glued together), basal portion of skull damaged. Rostrum equals braincase in length; both rostrum and braincase elongated; braincase moderately elevated above rostrum; upper premolars two rooted, much compressed laterally; both first and second upper premolars with anterior and posterior styles, anterior stronger (posterior style almost absent on first premolar); lower molars much compressed, with four cusps, talonid reduced, metaconid well developed and lingual to protoconid, others essentially in straight line.

Remarks.—The holotype was among a number of small mammals sent by the Genoa museum to Thomas for examination. The collector's name was twice spelled "Esposto" in Thomas' description. The name *genovensium* is a tribute to Thomas' Genoese friends at the Museo Civico, Genoa.

Choeroniscus

Choeronycteris Godmani Thomas

1903.   Ann. Mag. Nat. Hist., ser. 7, 11:288. March.

BMNH 79.12.24.1: adult male; in alcohol, skull removed; Guatemala; G. C. Champion; date of capture not specified. Holotype.

Skin.—Slightly faded. Nose leaf short, pointed.

Skull.—Rostrum narrow, shallow, approximately two-thirds length of braincase; braincase much expanded; palate long, posterior margin even with anterior margin of glenoid fossa; pterygoids expanded laterally, in contact with tympanum; upper incisors small, I1 peglike, I2 conical and vertical length greater than that of I1; upper canine long, laniary, with small posterior cingular style (style rather than shelf); upper premolars two in number, two rooted, triconate (anterior and posterior style small but distinct, about equal in size, central cusp much larger than styles and situated anterior to center of long axis of tooth); upper molars apparently three rooted; protocone and metacone distinct, paracone missing (if present, parastyle missing), metastyle present but weakly developed, central depression moderately well developed; first upper premolar separated from canine by a space equal to the anteroposterior length of first upper premolar, each succeeding tooth separated from preceding tooth by space approximately one-third the anteroposterior length of preceding tooth; lower premolars triconate, central cone more or less in center of long axis of tooth, only slightly higher than anterior and posterior styles; anterior and posterior styles well developed, almost equal in size to central cusp; premolars much compressed laterally, paraconid weakly developed and near midline of tooth, metaconid largest of three main cusps; talonid well developed, with two cusps and rather well-formed basin.

Remarks.—The name *Choeronycteris godmani* is a patronym for F. DuCane Godman, who presented the specimen to the British Museum.

Choeronycteris inca Thomas

1912.   Ann. Mag. Nat. Hist., ser. 8, 10:403. October.

BMNH 12.9.5.2: adult female; skin and skull; [Río] Yahuarmayo, 1200 ft., Puno, Perú; H. & C. Watkins, no. 1389, measured 64, 9.5, 19 [including tibia], 12; 7 February 1912. Holotype.

Skin.—Dorsal pelage dark reddish brown, basal two-thirds of individual hairs gray to pale reddish brown; ventral pelage paler than dorsal (basal two-thirds of individual hairs paler than distal portion and tending toward gray); membranes and ears dark reddish brown to blackish.

Skull.—Occiput and basal portion missing. Rostrum shallow and fairly narrow, two-thirds to three-fourths length of braincase; braincase evenly expanded and evenly rounded dorsally; no palatal notches posterior to zygomatic process of maxillary; palate extends posterior at least to anterior border of glenoid fossa; posterolateral expansion of pterygoid in contact with tympanum (tympanum missing from holotype, but there is no doubt that pterygoid was in contact with it); I2 slightly larger than I1; I2 separated from canine by space equal to approximately one-half greatest diameter of I2; upper canine with small anterointernal cingular style and small posterior cingular style; upper premolars two rooted; first upper premolar with main cusp slightly anterior to center of tooth, anterior and posterior styles present, space between canine and first premolar equal to two times the linear dimension of first upper premolar; second upper premolar like first but less compressed laterally, slightly larger than first; upper molars apparently three rooted, protocone and metacone distinct and well defined, paracone may be missing (if present, parastyle missing), central depression moderately defined; M1 and M2 equal in size; M3 smaller than M2; lower incisors absent; lower canine without cingular styles or shelf; lower premolars three in number (first slightly smaller than second, second and third equal in size, all three separated from each other by a small space), with primary cusp in center of tooth, anterior and posterior styles about as well developed as central cusp; first lower premolar close to (but not in contact with) canine; lower molars with five cusps, paraconid small (situated along midline of tooth), talonid well developed and depression well defined; m1 largest, m2 and m3 progressively smaller, separated by narrow space.

Remarks.—The name *Choeronycteris inca* was based on a single specimen, which obviously is the one reported here although Thomas recorded the collectors' field number as 709. The Río Yahuarmayo (sometimes spelled Yaguarmayo) is a small river in the Sahuaca district of the department of Puno; it joins the Río Inambari at 13°18′S, 70°17′W. Compared with the holotype of *Choeronycteris godmani*, that of *C. inca* is noticeably larger, with broader, deeper rostrum and larger teeth.

Choeronycteris minor Peters

1868.   Monatsb. Kön. preuss. Akad. Wiss. Berlin, p. 366.

SMNS 441: adult male; skin and skull; Surinam; A. Kappler; 1851. Holotype.

Skin.—Faded. Nose leaf short, broad; uropatagium broad, naked.

Skull.—Damaged. Skull and teeth as in TCWC 11169 female (a specimen of *Choeroniscus godmani*) except larger; rostrum and braincase noticeably broader; end of rostrum missing (canines and incisors missing from immediately in front of first upper premolars, but rostrum as long as in TCWC 11169).

Remarks.—Peters examined a single specimen, then, as now, housed in the Stuttgart museum.

CAROLLIINAE

Carollia

Carollia azteca Saussure

1860.   Rev. Mag. Zool. Paris, ser. 2, 12:480.

ZMB 2647: adult female; skin, skull not removed; México; Saussure; date of capture not specified. Syntype.

Skin.—Faded somewhat. Dorsal pelage banded, as in *Carollia brevicauda,* terminal band reddish brown; proximal two-thirds of forearm hairy; toes hairy.
Skull.—Second lower incisor visible from above.

Remarks.—Pine (1972:32-33) concluded that this specimen is referable to Saussure's [*Carollia azteca*] *Variété* (p. 482), a synonym of *Carollia brevicauda.* Berlin museum records seen by us indicate only that it was received as a type specimen of *Phyllostoma aztecum,* as it is labeled.

Carollia verrucata Gray

1844.   The zoology of the voyage of H.M.S. Sulphur . . . , 1(Mammalia):20, pl. 18, fig. 3.

BMNH 106.a: adult of undetermined sex; skin and partial skull; locality, collector, and date of capture unknown. Holotype.

Skin.—Dorsal pelage brownish, with broad, dark basal band; venter paler than above; forearms and toes hairy, as in TCWC 12068.
Skull.—Lower jaw V-shaped (not bowed); i2 not visible from above.

Remarks.—A notation on the label, presumably written by Gray, identified this specimen by the name *Arctibeus verrucatus,* a *nomen nudum* (Gray, 1843: 19); the author of another notation, "*Phyllostoma soricina* Lorn; *Hemidera perspicillatum,*" is unknown to us.

Glossophaga amplexicauda É. Geoffroy St.-Hilaire

1818.   Mem. Mus. Hist. Nat. Paris, 4:418, pl. 18, fig. A.

MNHN A.291: juvenile male (eight to 10 weeks of age); skin and partial skull; Rio de Janeiro, Brazil; Delalande fils and Auguste de St.-Hilaire; date of capture not specified. Syntype.

Skin.—Mounted and faded. Dorsal pelage with traces of three color bands, now pale; forearms hairy on proximal one-half; toes hairy. The bones of the forearms were removed in preparation, and it is not possible to determine how hairy the forearms were (some skin is missing), also the skin of the forearm is wrinkled, obscuring the exact position of the elbow.
Skull.—Recently removed, only rostrum and mandible remain. The second lower incisor visible from above; mandibular rami V-shaped but bowed outward; compared with TCWC 12090 and 12036, MNHN A.291 resembles 12090 more closely than it does 12036.

Remarks.—The condition of the skin and skull is poor, but the specimen obviously represents *Carollia,* not *Glossophaga* (É. Geoffroy St.-Hilaire's fig. A, pl. 18, is not that of any Paris museum specimen labeled *G. amplexicauda*). The catalogue number A.291 dates from *circa* 1868, at which time the specimen was recatalogued. "*Glossophaga soricina* Pallas / *Gloss amplexicaudata* Geoff. St. H. type / A.291," is written on the bottom of the block of wood to which the specimen is attached—the number and preceding portion of the notation were written at different times. For some reason, Geoffroy's *Glossophaga amplexicauda* was routinely misspelled by subsequent authors, perhaps because it was confused with *Glossophaga amplexicaudata* Spix, 1823, a synonym of *Glossophaga soricina* (Pallas). See Pine (1972:45-47) for a more detailed discussion of the name.

Two additional specimens (numbers 935, adult male, and 936, adult of undetermined sex), both *Carollia perspicillata*, are labeled *Glossophaga amplexicaudata* (emended to "*Carollia brevicauda*") and once were glued to boards for display in the public gallery. The numbers 935 and 936 belong to a cadre series used *circa* 1864; the catalogue numbers corresponding to these cadre numbers are 986 and 985, respectively (there are catalogue notations that 985=266= cadre 33 and that 986=A.267). Although catalogue entries, dated 1822, for numbers 985 and 986 indicate only that the two specimens came from A. St.-Hilaire, they appear to belong to a collection made jointly by Delalande fils and Auguste de St.-Hilaire (see statement by É. Geoffroy St.-Hilaire, 1818a:411, as well as that on p. 418). The date 1822 accompanies other specimens obtained by Delalande fils and A. St.-Hilaire and must refer to the date that these specimens were registered or formally received by the Paris museum. We consider the two specimens numbered 935 and 936 to be syntypes of *Glossophaga amplexicauda.*

Phyllostoma bicolor Wagner

1840.   Die Säugthiere . . . von Schreber, Supplementband, 1:400.

ZSM 126: adult of undetermined sex; skin, skull not removed; Brazil; collector and date of capture unknown. Holotype or syntype.

Skin.—Faded; some loss of hair on lower back. Dorsal pelage long and dense, with basal band reddish brown (in some areas there appears to be a very narrow band of white at base of hair), median band white (broadest band), distal band reddish brown; venter paler than dorsum, hairs with basal band of reddish brown, median band cream, distal band slightly darker than median; forearms hairy. Compared with TCWC 9932, 12036, 12068, 12090, 16424, and 16460, ZSM 126 resembles most closely TCWC 12090.

Skull.—In skin, apparently entire. First lower incisors missing, i2 visible from above.

Remarks.—An old green label affixed to ZSM 126 bears the inscription "*Phyllostoma bicolor* Wagner / (*Vampyrus soricinus* Spix) / Bras. / *Ph. brevicaudum* Wied" (Wied's name is written in pencil).

Phyllostoma calcaratum Wagner

1843.   Arch. Naturgesch., 1:366.

ZSM 141: adult female; skin; Brazil; collector and date of capture unknown. Holotype.

Skin.—Faded somewhat. Dorsal pelage short (approximately 7 millimeters) on shoulder and base of neck; color medium reddish brown (basal and distal bands reddish brown, median band pale gray white; ventral color paler, and grayer than above, banding indistinct; forearms moderately hairy (more so than in TCWC 16460 but less hairy than in TCWC 12090); toes as hairy as in TCWC 16460; nose leaf similar to that of TCWC 16460 (more rounded at tip, tapering less acutely from base than in TCWC 9932, 12036, 12068, 12090, and 16501).

Skull.—Not located and presumed lost.

Remarks.—The notation "*Phyllostoma calcaratum* Wagn / 1843 Brandt / Bras. / *Ph. brevicaudum*" (the last line written in pencil) appears on an old green label. The oldest label with the specimen is inscribed "J. G. W. Brandt, Hamburg / *Phyllostoma brachyotum* / WZ / Brasil." Wagner's statement (p. 367) that this specimen was from Brazil was based on information from Brandt, a Hamburg dealer in natural history specimens, from whom Wagner apparently obtained the holotype of *P. calcaratum*. The meaning of Wagner's (p. 367) statement "Mus. Monac." is unknown to us.

R[hinops]. minor Gray

1866.   Proc. Zool. Soc., London, p. 115.

BMNH 49.10.15.13: juvenile male (approximately eight weeks old); skin and partial skull; Bahia, [Brazil]; Castelnau; date of capture not specified. Holotype or syntype.

Skin.—Dorsal pelage brown (basal band dark gray, median band pale cream-brown, distal band brown); forearms and toes hairy.

[Vespertilio] perspicillatus Linnaeus

1758.   Systema naturae . . . , ed. 10, 1:31.

BMNH 67.4.12.597: female, probably subadult; in alcohol, skull removed; [Surinam (Thomas, 1911:130)]; collector and date of capture unknown. Holotype.

Skin.—Condition fair, pelage faded; bones moderately soft. Forearm without hair.

Skull.—Rostrum and mandible only. Second lower incisor visible from above; mandible V-shaped.

Remarks.—This specimen is from the Lidth de Jeude collection (no. 597), as noted on the skull label, and Thomas (1892) concluded that it represents Seba's original specimen (no. 2). The specimen was registered in the British Museum collection as an adult, but probably it is a subadult. See Pine (1972) for a detailed history of the name *Vespertilio perspicillatus* and additional comments on its holotype.

Rhinophylla

Rhinophylla pumilio Peters

1865.   Monatsb. Kön. preuss. Akad. Wiss. Berlin, p. 355.

ZMB 3060: adult female; in alcohol, skull removed; Brazil; Ruppell; date of capture not specified. Holotype.

Skin.—Faded, otherwise condition good. Interfemoral membrane without fringe of long hairs.

Skull.—Occiput broken.

Remarks.—See Carter (1966) for additional information on this specimen.

Rhynophylla cumilis Kappler

1881.   Hollandisch-Guiana, p. 163.

SMNS 289(1): adult female; in alcohol, skull removed; Surinam; A. Kappler; 1845/1860. Not a type.

SMNS 289(2): juvenile female; in alcohol, skull removed; Surinam; A. Kappler; 1845/1860. Not a type.

ZMB 3346: adult female; in alcohol; Surinam; A. Kappler; date of capture not specified. Not a type.

Remarks.—All three specimens are referable to *Rhinophylla pumilio* Peters, each having trilobed lower incisors and lacking a fringe of hair on the free edge of the interfemoral membrane. The name "*Rhynophylla cumilis* Peters" (p. 163) appeared only in a list of Surinam specimens in the Stuttgart museum, is an obvious incorrect spelling of *Rhinophylla pumilio* Peters, and is unavailable.

STENODERMINAE

Sturnira

Corvira bidens Thomas

1915.   Ann. Mag. Nat. Hist., ser. 8, 16:311. October.

BMNH 15.7.11.7: adult male; skin and skull; Baeza, upper Río Coca, 6500 feet, Napo, Ecuador; Walter Goodfellow, no. 19; April 1914. Holotype.

Skin.—Dorsal pelage dark gray, basal band gray, median band pale buff, distal band dark gray; venter reddish gray, pelage with basal band gray, distal band reddish gray; membranes and ears blackish.

Skull.—Rostrum with supraorbital region inflated; zygomata weak; basisphenoid pits with strong median ridge; first upper incisors converge distally; I1 blunt, with posterolateral style; cheekteeth not in contact with each other; i1 bilobed; lower canine with broad posterior style; lower cheekteeth not in contact.

Remarks.—Although Thomas indicated that this specimen was "somewhat immature," it is an adult.

Phyllostoma albescens Wagner

1847.   Abh. Bayerischen Akad. Wiss., 5:177.

ZSM 129: juvenile female; skin, skull not removed; Ypanema [Ipanema], São Paulo, Brazil; Johann Natterer, no. 26; 8 January 1822. Holotype.

Skin.—Faded; that of a juvenile old enough to fly, probably six to 10 weeks old.

Skull.—Incomplete, most of braincase missing. Lower incisors trilobed, right i1 scarcely so.

Remarks.—The original label for ZSM 129 bears the notation "No. 26 / 8 Janer 822 / Foemine;" an old green label is inscribed "*Phyllostoma albescens*

Wagn / ♀ / E. Mus Vindal / 1840 / Ypanema / (Brazil)." Purportedly (Pelzeln, 1883), a single specimen of *Phyllostoma albescens* was collected at Ypanema, which Joseph Natterer apparently sent to Wagner for examination (correspondence on file in Vienna museum); also, Wagner, in his description of *P. albescens,* referred to a single specimen. ZSM 129 seems to be referable to *Sturnira lilium* (É. Geoffroy St.-Hilaire, 1810).

An unnumbered, spirit-preserved juvenile male (labeled by Fitzinger *circa* 1850 "*Nyctiplanus albescens* Wagner / Brasilia / Br. / Adult") in the Vienna museum appears to be referable to *Sturnira ludovici* and would seem to have no type status; Johann Natterer collected nine specimens of *Sturnira* (all presumed to be *Phyllostoma excisum* Wagner, 1842, apparently another synonym of *S. lilium*) from a tree hollow at Ypanema.

Phyllostoma excisum Wagner

1842.　Arch. Naturgesch., 8(1):358.

ZSM 137: adult male; skin, skull not removed; Rio de Janeiro, Brazil; Johann Natterer; 12 August 1818. Syntype.

Skin.—Faded. Some dark brown (with reddish cast) areas on dorsum; no indication of shoulder patches.
Skull.—Apparently entire. First upper incisor faintly bilobed, right I1 essentially entire on cutting edge and a little longer than the left I1; lower incisors trilobed.

Remarks.—A label written by Johann Natterer is inscribed "N. 26. Rio / v. 12 Aug. 18 / Mas." The notation "26 *Dyphilla ecaudata* Sp. [written by Joseph Natterer] / *Phyllostoma excisum* [written by Wagner in red ink]" appears on a second label. ZSM 137 is one of two specimens (the other in spirits) sent to Wagner by Joseph Natterer. The fluid-preserved specimen was not located and is presumed lost.

Phyllostoma fumarium Wagner

1847.　Abh. Bayerischen Akad. Wiss., 5:178.

ZSM 58: adult of undetermined sex; skin, skull not removed; possibly Brazil; collector and date of capture unknown. Holotype.

Skin.—Slightly faded. Dorsal pelage medium brown; venter paler than dorsum, with some gray.
Skull.—In skin; occiput and part of braincase missing. Lower incisors trilobed.

Remarks.—Wagner had a single specimen, which he believed to be from Brazil (p. 179), but the label was lost when the specimen was prepared as a skin from one originally preserved in spirits.

Phyllostoma lilium É. Geoffroy St.-Hilaire

1810.　Ann. Mus. Nat. Hist. Nat. Paris, 15:181.

MNHN (not numbered): adult male; skin and skull; Asunción, Paraguay; Felix de Azara; date of capture not specified. Holotype or syntype.

Skin.—Faded, otherwise in good condition. No evidence of shoulder patches.

Skull.—First upper incisor seems to have been almost pointed (one is broken away near the cutting edge, and the other may be chipped on the cutting edge); first and second lower incisors trilobed (i1 most obviously so).

Remarks.—The specimen is identified as type no. 195. According to É. Geoffroy St.-Hilaire, this specimen was taken by Azara (presumably at Asunción); the specimen is labeled simply as having come from "Amerique."

Phyllostoma oporophilum Tschudi

1844.   Untersuchungen uber die Fauna Peruana, p. 64, pl. 2.

ZSM 145: adult of undetermined sex; skin, skull not removed; Perú; collector and date of capture unknown. Holotype or syntype.

Skin.—Faded. Dorsal hair with three bands; ventral hair with two bands.

Skull.—Occipital region seems to be missing. Lower incisors bifid; I1 bifid.

Remarks.—Although usually placed in synonymy with *S. lilium*, this name does not belong there. *Sturnira oporophilum* is more closely related to *S. ludovici* than to *S. lilium* and probably is the valid name for some populations now referred to *S. ludovici*. A recent green label is inscribed "Mittlere Waldregionen zwischen 12 und 14 sudl. Breite."

Sturnira Spectrum Gray

1842.   Ann. Mag. Nat. Hist., 10:257. December.

BMNH 42.12.2.4: adult of undetermined sex; skin and skull; Brazil; Leadbetter; date of capture not specified. Holotype or syntype.

Skin.—Condition fair. Epaulet on shoulder.

Skull.—Badly damaged and uncleaned. Upper incisors converge distally, blunt; trilobed but weakly so on left side—perhaps due to wear.

Uroderma

Uroderma bilobatum Peters

1866.   Monatsb. Kön. preuss. Akad. Wiss. Berlin, p. 394.

ZMB 409: adult male; skin, skull not removed; Cayenne, French Guiana; Delbruk; date of capture not specified. Syntype.

ZMB 410: adult male; skin, skull not removed; Cayenne, French Guiana; Delbruk; date of capture not specified. Syntype.

ZMB 411: juvenile of undetermined sex; skin, skull not removed; São Paulo, Brazil; Sello; date of capture not specified. Syntype.

Skin.—The following applies to all three syntypes: skin somewhat faded; dorsal color gray with reddish cast, dorsal hairs paler at base; thin white line along midline of back; venter paler than dorsum; interfemoral membrane essentially naked above and below, margin without fringe of hair.

Skull.—None has been removed; M3 and m3 can be seen in ZMB 409 and 410.

Remarks.—ZMB 411 is entered in the catalogue as a "var. *personatus* Natterer." ZMB 409, 410, and 411 first were entered as *Artibeus (Uroderma) personatus*; that name is lined out, and below it is written "*Artibeus lineatus* Gervais *Ph. lineatum* Geoffr."; that name also is lined out and below it is written "*Uroderma bilobatum* Ptrs." These notations seem to have been made by Peters (see p. 395). ZMB 411 was deposited originally in the Berlin museum, but the other two were obtained from Rüppells at the Senckenberg museum in Frankfurt a.M. These two were preserved in fluid at the time Peters described them and noted their sex.

Uroderma Thomasi Andersen

1906.   Ann. Mag. Nat. Hist., ser. 7, 18:419. December.

BMNH 1.2.1.37: adult male; in alcohol, skull removed; Bellavista [15° S, 68° W], 1400 m. [in department of La Paz], Bolivia; Perry O. Simons, no. 1259 (presented by O. Thomas); 11 October 1900. Holotype.

Skin.—Not located and presumed lost.
Skull.—Damaged slightly. I1 bifid; i1-2 weakly bifid.

Remarks.—Andersen stated that a second specimen from Reyes, Bolivia, 13° S, 67° W was presented by G. Doria.

Vampyrops

Artibeus vittatus Peters

1860.   Monatsb. Kön. preuss. Akad. Wiss. Berlin, p. 225 (for 1859). 28 February.

ZMB 568: adult male; in alcohol, skull removed; Puerto Cabello, Carabobo, Venezuela; von Appun; date of capture not specified. Holotype or syntype.

Skin.—Some loss of hair. Pelage dark gray on dorsum, paler on venter; narrow white stripe on back from middle of shoulders to base of interfemoral membrane.
Skull.—First lower incisor missing, second lower incisor bilobed.

Phyllostoma lineatum É. Geoffroy St.-Hilaire

1810.   Ann. Mus. Nat. Hist. Nat. Paris, 15:180.

MNHN 953: adult of undetermined sex; skin and skull; Asunción, Paraguay; Felix de Azara; date of capture not specified. Holotype.

Skin.—Faded. Evidence of four well-defined facial stripes and median stripe on back.
Skull.—Lower incisors bilobed; skull much smaller than that of TCWC 12172 (*V. dorsalis,* male, from 1 mi. N Zaruma, 4600 ft., El Oro, Ecuador).

Remarks.—Label information indicates that this specimen is one of Azara's; therefore, it is probably from Asunción. Cabrera's (1958) statement regarding the type locality was based on this assumption; É. Geoffroy St.-Hilaire (1810: 186) stated only that the species (in this case, specimen) was from Paraguay. The type number is 194, but the specimen is identified as 953 on its label (933 in Rode, 1941, is an error). This specimen was once on exhibit in the public gallery. "Phyllostome raye" (the vernacular name applied by É. Geoffroy St.-Hilaire to this species) is written on the wing.

Vampyrops dorsalis Thomas

1900. Ann. Mag. Nat. Hist., ser. 7, 5:269. April.

BMNH 99.12.5.1: juvenile male; skin and skull; Paramba, 1100 m., Ecuador; R. Miketta, no. 61; 14 April 1899. Holotype.

Skin.—Faded somewhat. Hair on dorsum with narrow basal band pale, followed by broad band of pale brown (about one-half length of hair), band of pale buff, and distal band of brown; dorsal white line from shoulders to rump, along midline, well defined; facial stripe from nose leaf to ears pale buff; perhaps a faint stripe from corners of mouth to ears; hair on venter gray, with white frosting; forearm hairy on proximal two-thirds; interfemoral membrane narrow, with fringe of hairs; membranes, ears, and nose leaf gray brown; epiphyses not fused to shaft of metacarpals.

Skull.—Rather elongate in the interorbital region, otherwise like those of other Vampyrops; lower incisors faintly bilobed. Compared with TCWC 12172 (V. dorsalis, male, from 1 mi. N Zaruma, 4600 ft., El Oro, Ecuador).

Vampyrops Helleri Peters

1866. Monatsb. Kön. preuss. Akad. Wiss. Berlin, p. 392.

ZMB 3276: adult, probably female; skin, skull not removed; México; Heller; 1850. Syntype.

Skin.—Pelage reddish brown above, paler below; four facial stripes, those from nose leaf to ears broad, white, those from corner of mouth to ears not so broad, white; distinct white line along midline of back; margin of interfemoral membrane with dense fringe of hair; proximal half of forearm with dense hair.

Skull.—Most of cranium posterior to rostrum apparently removed.

Remarks.—Two specimens, both collected in México by Heller, were obtained from the Vienna museum. Both were mentioned by Peters, but only one (ZMB 3276) was found. There is no other Mexican specimen of Vampyrops helleri in the Berlin museum, and none from México was found in the Vienna museum.

Vampyrops lineatus sacrillus Thomas

1924. Ann. Mag. Nat. Hist., ser. 9, 13:236. February.

BMNH 23.12.12.9: adult male; skin and skull; Rio Doce, Espírito Santo, Brazil; Roberto Dabbene, no. 54, Buenos Ayres museum, measured ——, ——, 11, 15; date of capture not specified. Holotype.

Skin.—Dorsal pelage dull reddish brown, banded as in holotype for Vampyrops dorsalis; dorsal stripe from head to rump; narrow white facial stripes from nose leaf to ears and from corners of mouth to ears; ventral pelage with basal color cream, gray distally, white tips; lateral basal margin of ear with very narrow white border; membranes, ears, and nose leaf brown in color.

Skull.—Occiput broken and missing. Rostrum and interorbital region relatively broad and short; zygomata relatively weak.

Remarks.—This specimen was obtained from Dr. Roberto Dabbene (whose name appears on the specimen label) of the Buenos Ayres museum. Thomas stated that another specimen was from Lagoa Santa, Minas Gerais; presumably, it is BMNH 93.1.9.14 (K. F. Koopman, personal communication).

Vampyrops oratus Thomas

1914.   Ann. Mag. Nat. Hist., ser. 8, 14:411. November.

BMNH 14.7.27.1: adult male; skin and skull; Galifari [6500 ft.], Sierra del Avila, near Caracas [in Distrito Federal], Venezuela; S. M. Klages, no. 2, measured 72, 0, 11, 15; 13 December 1913. Holotype.

Skin.—Dorsal pelage like that of holotype for *Vampyrops dorsalis* except that distal band is darker; ventral pelage gray, with white frosting; membranes and nose leaf blackish, ears brownish, interfemoral membrane narrow, with fringe of hairs along free edge.

Remarks.—The locality on the specimen label was written as "Galipare, Cerro del Avilo, nr. Caracas."

Vampyrops recifinus Thomas

1901.   Ann. Mag. Nat. Hist., ser. 7, 8:192. September.

BMNH 81.2.16.4: adult male; in alcohol, skull removed; Pernambuco [Recife], Pernambuco, Brazil; W. A. Forbes; date of capture not specified. Holotype.

Skin.—Condition poor, hair without color and slipping.

Skull.—Rostrum and interorbital region relatively short and broad (as in TCWC 12172); rostrum not as broad as in holotype for *Vampyrops lineatus sacrillus*, but maxillary toothrow broader; zygomata relatively strong; I1 with three lobes; lower incisors trilobed.

Remarks.—Sanborn (1955) erroneously reported the British Museum registry number to be 81.3.16.4, as did Thomas.

Vampyrops zarhinus incarum Thomas

1912.   Ann. Mag. Nat. Hist., ser. 8, 9:408. April.

BMNH 12.1.15.1: adult male; in alcohol, skull removed; Pozuzo, Perú; L. Egg; date of capture not specified. Holotype.

Skin.—Faded to the extent that it is difficult to discern color and stripes.

Skull.—Damaged (cracked). Noticeably smaller than most of the skulls of *Vampyrops zarhinus* (= *V. helleri*) in British Museum collection; rostrum relatively long, interorbital region relatively narrow; I1 with faint indication of three lobes; toothrow relatively narrow; lower incisors trilobed.

Remarks.—See Rouk and Carter (1972) for a discussion of the names *V. zarhinus* and *V. zarhinus incarum*.

Vampyrodes

Vampyrodes ornatus Thomas

1924.   Ann. Mag. Nat. Hist., ser. 9, 13:532. May.

BMNH 24.3.1.63: adult female; skin and skull; San Lorenzo, 500 ft., Río Marañon, nearly opposite mouth of [Río] Huallaga [in department of Loreto], Perú; Latham Rutter, no. 426; 13 March 1923. Holotype.

Skin.—Dorsal pelage brown, white stripe from head to rump; four facial stripes, dorsal stripes from nose leaf to back of ears; most of ear bordered with white; interfemoral membrane fringed with short stiff hairs.

Skull.—I1 more pointed than in holotype for *Vampyrodes caracciolae*; lower incisors bilobed.

Remarks.—Thomas gave 13 November as the date of capture. He also indicated that there were four paratypes (none identified by number), one of which was from Masisea, "1000 ft.," on the Río Ucayali.

Vampyrops Caracciolae Thomas

1889. Ann. Mag. Nat. Hist., ser. 6, 4:167. August.

BMNH 89.6.10.2: juvenile of undetermined sex; skin and skull; Trinidad; H. Caracciolo, no. 2; date of capture not specified. Holotype.

Skin.—Dorsal stripe; four facial stripes (stripe from nose leaf to back of ear on top of head, and one from corner of mouth to base of ear; most of ear bordered with yellow; interfemoral membrane fringed with hair.

Skull.—Lower incisors bilobed.

Remarks.—Mr. H. Caracciolo, of Trinidad, sent this specimen, along with a few more of other species, to Thomas, who, thinking that the collector's name was spelled "Caracciola," originally proposed the species name "*caracciolae*." Thomas later corrected the spelling of Caracciola to Caracciolo and, accordingly, emended the species name to *caraccioli* in the British Museum's set of Thomas' separates. Thereafter, Thomas (1893, 1920, 1924) used the emended name to refer to this species. Cabrera (1958:82), without explanation, changed the ending of this name to *caraccioloi*; following Cabrera (*loc. cit.*), Goodwin and Greenhall (1961:257) argued that the correct form of the name should have been written *caraccioloi* and that Thomas' (1893) use of *caraccioli* must have been a typographical error. It was not, of course, because the correct Latin genetive singular case for the name Caracciolo (derived from the Latin nominative *Caracciolus*) would be Caraccioli—originally, Thomas obviously treated the Italian name that he thought to be Caracciola as though it were Latin, for which the correct genetive singular case would be Caracciolae. The specific epithet cannot be spelled *caraccioloi* because it is not the genetive singular of the name Caracciolo; we consider the spelling of this name to be a simple matter of grammar and, following recommendation 31A of the International Code for Zoological Nomenclature (1963), encourage the use of the name *caraccioli*.

Vampyressa

Chiroderma bidens Dobson

1878. Catalogue of Chiroptera in the collection of the British Museum, p. 535.

BMNH 69.3.31.12: adult female; skin and skull; Santa Cruz, Río Huallaga [in department of Loreto], Perú; E. Bartlett; 7 August 1868. Holotype.

Skin.—Dorsum with stripe along midline of back; facial stripes four in number, white; ears and nose leaf bordered with yellow; tragus yellow; interfemoral membrane naked, without fringe.

Skull.—Upper incisors missing; first lower incisor bifid.

Remarks.—Dobson examined a single specimen.

Phyllostoma pusillum Wagner

1843.   Arch. Naturgesch., 9(1):366.

ZSM 1843/2: probably juvenile, male; skin, skull not removed; Sapitiva [in state of Rio de Janeiro], Brazil; Johann Natterer; 6 August 1818. Holotype.

Skin.—Faded somewhat. Dorsal pelage pale brown, with slight reddish cast, basal band brownish, median band pale buff, distal band pale brown (paler and with more red than basal band); venter paler than dorsum, with some gray; facial stripes distinct, white, four in number; interfemoral membrane narrow, some hair on dorsal surface, median ventral surface hairy, margin without fringe except on median portion; hair on back extends beyond margin of interfemoral membrane; forearms with dense hair on proximal one-half.

Skull.—Occiput apparently missing. Lower incisors two in number, bifid; I1 weakly bifid; molars apparently 2/2.

Remarks.—Four labels are affixed to the specimen; they are inscribed as follows (in chronological order, the oldest being first): "N.16 Sapitiva / v.6. Aguit 1818" (probably Johann Natterer's original label); "19 / *Phyllostoma pusillum* Natt." (probably affixed by Wagner); "*Phyllostoma pusillum* Natt. ♂ juv. / 1844. / Sapativa, Bras. / Nr.: 1843 / 2." (green label); "*Phyllostoma pusillum* (Natterer) in Wagner, 1843, *in:* Diagnosen neuer Arten brasilischer Handflugler. Arch. Naturgesch, Berlin, 9,1,355,1843. Von A. Wagner" (face of type label) and back "S. Auch: A. Wagner 'Beitrage zur Kenntnis der Saugetiere Amerikas'. Abh. math. phys. Klasse Bayer. Akad. Wiss., 5, Abt. 1, p. 53 u. Taf II (p. 89), 1847." Wagner, according to an invoice in the Vienna museum files, examined a single specimen. Cabrera (1958:83) erred in citing the original description of *P. pusillum.*

Vampyressa melissa Thomas

1926.   Ann. Mag. Nat. Hist., ser. 9, 18:157. July.

BMNH 26.5.3.4: adult female; skin and skull; Puca Tambo [approximately 6°9'S, 77°16'W, 1480 m., municipality of Chachapoyas, department of Amazonas], 5100 ft., Perú; R. H. Hendee (Godman-Thomas Expedition), no. 408, measured 54, —, 10.5, 16; 15 January 1926. Holotype.

Skin.—Dorsal pelage soft and thick, approximately 8.5 in length, color brown, with basal band gray, median band pale buff, distal band dull brown; four facial stripes moderately defined; ventral pelage pale gray, basal band gray; ear yellow on basolateral border; tragus yellow; wing membranes blackish; interfemoral membrane brown; legs hairy; interfemoral membrane furred on ventral surface, free edge with fringe of hair.

Skull.—First upper incisor bifid; i1-2 bifid.

Remarks.—The notation "caught in house in early evening" appears on the specimen label. Thomas apparently examined a single specimen.

Vampyressa nattereri Goodwin

1963.   Amer. Mus. Novit., 2125:16.

RNH 17256: adult male; skin and skull; Ipanema [Ypanema], São Paulo, Brazil; Johann Natterer, no. 19; 18 August 1819. Holotype.

Skin.—Faded.
Skull.—Entire.

Remarks.—The notation "Brazil / Natterer / Jent. 1888, p. 209, no. *c*" appears on the specimen label. The date and author of this inscription are unknown to us and could have been made subsequent to Goodwin's (1963:16) statement that the specimen probably was taken at Ipanema, Brazil, on 18 August 1819 by Johann Natterer, based on a species record card of Johann Natterer's Brazilian collection in the Naturhistorisches Museum, Vienna. This card shows that a male identified as *P. pusillum* Wagner, collected by Natterer at Ypanema in 1819, was given to Temminck at the Musée d'Histoire Naturelle de sa Majesté le Roi des Pays-Bas, now known as the Rijksmuseum van Natuurlikje Historie, Leiden. The name *Vampyressa nattereri* first appears on p. 14 as a figure legend.

Vampyressa nymphaea Thomas

1909.   Ann. Mag. Nat. Hist., ser. 8, 4:230. September.

BMNH 9.7.17.40: adult male; skin and skull; Novita, 150 ft., Río San Juan, Chocó, Colombia; M. G. Palmer, no. 185, measured 59, ——, 10, 16; 28 November 1908. Holotype.

Skin.—Dorsal pelage gray, hairs scarcely banded gray, paler gray, gray; facial stripes four in number, well defined, broad and white; slight trace of black stripe on rump; legs not hairy; interfemoral membrane without long hairs, ventral surface without fringe of hair on free edge.
Skull.—Incisors missing.

Remarks.—Thomas reported the collector's number to be 135, apparently a typographical error. He examined no other specimens.

Vampyressa thyone Thomas

1909.   Ann. Mag. Nat. Hist., ser. 8, 4:231. September.

BMNH 97.11.7.77: adult male; in alcohol, skull removed, Chimbo, 1000 ft., near Guayaquil, Bolívar, Ecuador; W. Rosenberg; 30 April 1897. Holotype.

Skull.—First upper incisor bifid; lower incisors weakly bifid.

Remarks.—Thomas also examined four other specimens from Chocó, western Colombia, but did not identify them by number or specific locality.

Vampyressa venilla Thomas

1924.   Ann. Mag. Nat. Hist., ser. 9, 13:533. May.

BMNH 24.3.1.73: adult female; skin and skull; San Lorenzo, 500 ft., Río Marañon [in department of Loreto], Perú; Latham Rutter, no. 417, measured 43, ——, ——, 14; 12 November 1923. Holotype.

Skin.—Dorsal pelage pale buff, with basal band pale buff, median band very pale buff or cream, distal band pale buff, four white facial stripes, no mid-dorsal stripe; ventral pelage dull pale buff, hairs banded as on dorsum; membranes blackish; ears yellowish brown, border cream; legs not hairy; interfemoral membrane with hair on ventral surface, free edge with sparce fringe of hair.

Skull.—First upper incisor bifid; lower incisors weakly trifid.

Remarks.—"Caught flying near house in evening" is noted on the specimen label. Thomas examined a single specimen.

Chiroderma

Ch[iroderma]. doriae Thomas

1891.   Ann. Mus. Civ. Stor. Nat. Giacomo Doria, ser. 2, 10:881. 25 June 1891.

BMNH 44.9.2.6/49.8.16.29: adult, probably female; skin and partial skull; Minas Gerais, Brazil; Parzudaki; date of capture not specified. Holotype.

Skin.—Dorsal stripe apparently lacking; facial stripes four in number, pale; forearms and dorsal surface of interfemoral membrane hairy.

Skull.—Crowns of I1 converge, medial surfaces parallel; i1-2 small, weakly trifid.

Remarks.—The skin is numbered 44.9.2.6; the skull, 49.8.16.29 and 1276.a. The skull label bears the notation "*Ch. villosum* (Fig. Pl. xxix. fig. 2) *Nec.* Peters." This specimen is the one described in some detail and figured by Dobson (1878) as *Chiroderma villosum.*

The name is a patronym for the Marquis Giacomo Doria, a naturalist at Genoa and patron of the Museo Civico di Storia Naturale di Genova. A single specimen was examined.

Chiroderma salvini Dobson

1878.   Catalogue of the Chiroptera in the collection of the British Museum, p. 532, pl. 29, figs. 3-3b.

BMNH 68.8.16.2: adult male; in alcohol, skull removed; Costa Rica; Osbert Salvin; date of capture not specified. Holotype.

Skin.—Faded badly.

Skull.—First upper incisors convergent (crowns convergent almost to tip), roots noticeably separated; lower incisors bifid.

Remarks.—The number 1953a and the notation "Cat. Chir. pl. xxix. fig. 3" appears on the skull label. Dobson examined a single specimen.

Chiroderma villosum Peters

1860.   Monatsb. Kön. preuss. Akad. Wiss. Berlin, p. 748.

ZMB 408: adult female; skin, skull not removed; Brazil; Sello; date of capture not specified. Syntype.

Skin.—Faded. Pelage yellow brown on dorsum, median stripe lacking; paler on venter; facial stripes two in number, indistinct, from nose leaf to ears; legs and dorsal surface of interfemoral membrane with moderately long hair, but free edge of membrane essentially naked.

Skull.—Upper incisors parallel, not convergent as in *Chiroderma salvini*; i1 (one of pair is missing) faintly bilobed; i2 peglike, crown evenly rounded.

Remarks.—The label indicates that ZMB 408 came from Brazil, but no specific locality is given; Cabrera's (1958:85) use of Venezuela for the type locality is undocumented. Peters examined a partial skeleton from the "anatomischen Museum," in addition to the skin listed above, but we did not attempt to locate it. Peters stated (p. 754) that the female (ZMB 408) came from Brazil, probably from the Sello collection; the partial skeleton had been in the anatomical museum for many years, and its origin was unknown.

The name *Chiroderma villosum* first appears on p. 747, as the type species for *Chiroderma*, but the description of this species begins on p. 748 and continues to p. 754.

Mesophylla

Mesophylla Macconnelli Thomas

1901.   Ann. Mag. Nat. Hist., ser. 7, 8:145. August.

BMNH 1.6.4.64: adult female; skin and skull; Kanuku Mountains, British Guiana [Guyana]; J. J. Quelch (presented by F. V. McConnell), no. 155, measured 45, ——, ——, 12.5, forearm 30, expanse 241; 9 November 1900. Holotype.

Skin.—Dorsum almost white on head and shoulders, becoming dull buff on rump; venter almost white; membranes blackish, bones yellowish; ears and nose leaf yellow; interfemoral membrane yellowish brown.

Remarks.—The collector entered the notation "rocks" on the skin label.

Artibeus

Artibeus carpolegus Gosse

1851.   A naturalist's sojourn in Jamaica, p. 271.

BMNH 47.12.27.13: adult male; in alcohol, skull removed; Content, Jamaica; P. H. Gosse; date of capture not specified. Syntype.

Skin.—Faded.
Skull.—Molars 2/3.

Remarks.—The British Museum apparently received two Jamaican specimens (a male and a female) from Gosse, but only the male (labeled type) was found. Gosse gave Content as the locality for the specimens upon which he based the name *Artibeus carpolegus*, but only Jamaica appears on the label attached to the specimen jar; according to Dobson (1878:520), only the female was reported as having come from Content.

Artibeus cinereus bogatensis Andersen

1906.   Ann. Mag. Nat. Hist., ser. 7, 18:421. December.

BMNH 99.11.4.35: adult male; skin and skull; Curiche, Bogotá [in department of Cundinamarca], Colombia; G. D. Child, no. 10 III, measured ear, 15; 16 August 1895. Holotype.

Skin.—Facial stripes four, moderately developed; ears with narrow white border on lateral and medial basal portion; tragus with anterior border cream; interfemoral membrane with long hair on central region of ventral surface, fringe of hair on central portion of free edge.

Skull.—Molars 2/2; i1-2 bilobed; rostrofacial angle approximately 140°, face long and evenly sloped.

Remarks.—Andersen examined a total of nine specimens with skulls but identified only the type by number.

Artibeus concolor Peters

1865.   Monatsb. Kön. preuss. Akad. Wiss. Berlin, p. 357.

ZMB 2617: adult male; in alcohol; skull removed; Paramaribo, Surinam; aus Hamburg; date of capture not specified. Holotype or syntype.

Skin.—Faded. Interfemoral membrane with moderately long hair on all but extreme distal edge of dorsal surface, ventral surface of interfemoral membrane with short hair on median proximal portion only; legs with moderately long hair; feet less hirsute than legs.

Skull.—Molars 3/3; lower incisors bilobed.

Artibeus fallax Peters

1865.   Monatsb. Kön. preuss. Akad. Wiss. Berlin, p. 355.

RNH 13083: adult female; skin, skull removed; Surinam; Dieperink; date of capture not specified. Lectotype.

Skin.—Faded, more so ventrally than dorsally. Dorsal pelage gray, hair banded as in TCWC 12319 (*Artibeus lituratus* from Popayán, Cauca, Colombia); ventral pelage now reddish gray with slight frosting.

Skull.—Molars 3/3, otherwise RNH 13083 similar to TCWC 12319 except that interorbital constriction of RNH 13083 broader and postorbital process not as well developed.

ZMB 566: adult female; in alcohol, skull removed; British Guiana [Guyana]; Schomburgk; date of capture not specified. Paralectotype.

Skin.—Faded badly. Interfemoral membrane with short hair on proximal three-fourths of dorsal surface; legs with very fine, short hairs, feet moderately hairy. Hair on legs shorter than in TCWC 12319; more like that of TCWC 12334 and 12327.

Skull.—Molars 3/3; posterior pair weakly developed. This syntype seems to be of the same species as the lectotype (Leiden museum) of *A. fallax.* Skull of ZMB 566 more closely resembles that of TCWC 12327 and 12334 than any of the others with which it was compared, but is a little smaller than either of these; supraorbital and postorbital regions much like those of 12327 and 12334. The skull of ZMB 566 is not like that of TCWC 12319 (*A. lituratus* from Popayán, Cauca, Colombia).

Remarks.—Apparently ZMB 566 long has been considered a type specimen of *Artibeus fallax* and may well have been labeled as such by Peters. We found no reason to doubt that ZMB 566 is the "weibliches Exemplar dieser Art in

Wiengeist aus Guiana" referred to by Peters (p. 357). Peters had a single specimen at the Berlin museum, but stated that "andere trockene Exemplare befinden sich im Reichsmuseum zu Leiden aus Surinam." Husson (1962:175) selected as lectotype one (RNH 13083) of three Leiden specimens apparently seen by Peters. The other paralectotypes (13081 and 13082) at Leiden are sub-adults. In designating a lectotype, Husson also restricted the type locality to Surinam. Cabrera (1958) earlier restricted the type locality to Cayenne, but Husson argued that there is no evidence that Peters saw a specimen from that place. Indeed, Peters mentioned only the Berlin specimen from Guiana ("Br. Guiana" on the specimen label) and other dry specimens from Surinam. Although we believe a valid argument could be made for recognizing the British Guianan specimen as a holotype, the situation probably is best left as it stands.

Artibeus glaucus Thomas

1893.   Proc. Zool. Soc., London, p. 336, pl. 29, figs. 7-9.

BMNH 94.8.6.13: juvenile female; in alcohol, skull removed; Chanchamayo [in departament of Junín], Perú; Kalinowski; date of capture not specified. Holotype.

Skin.—Condition very poor; it is impossible to discern any attributes of the pelage.
Skull.—Molars 2/3; facial outline as in *Artibeus watsoni* (rostrofacial angle approximately 150°, facial slope short); skull in general very much like that of TCWC 12281.

Remarks.—Thomas examined a single specimen.

Artibeus Jamaicensis Leach

1821.   Trans. Linn. Soc., London, 13:75.

BMNH (not numbered): adult of undetermined sex, probably female; skin; Jamaica; J. S. Redman; date of capture not specified. Holotype.

Skin.—Dorsum gray brown; facial stripes not apparent; venter yellowish, frosted with pale yellow.
Skull.—Not located and presumed lost.

Remarks.—Although there is no locality data with the specimen, Leach stated that it came from Jamaica. Leach apparently examined a single specimen, then housed in the Brookes Museum.

Artibeus jamaicensis aequatorialis Andersen

1906.   Ann. Mag. Nat. Hist., ser. 7, 18:421. December.

BMNH 0.2.9.13: adult male; skin and skull; Zaruma, 1000 m. [in departament of El Oro], Ecuador; Perry O. Simons, no. 395, measured 88, 0, 17, (23?); 17 June 1899. Holotype.

Skin.—Dorsal pelage dark gray with a slight brownish cast; ventral pelage gray with white frosting; wings tipped in white; general appearance very much like that of TCWC 12282 (*A. j. aequatorialis* from 15 mi. N Catacocha, 2000 ft., Loja, Ecuador).
Skull.—Molars 2/3.

Remarks.—Andersen examined a total of nine specimens, including eight skulls, but designated only the holotype by number.

Artibeus nanus Andersen

1906.   Ann. Mag. Nat. Hist., ser. 7, 18:423. December.

BMNH 89.1.30.5: adult male; in alcohol, skull removed; Tierra Colorado, Sierra Madre del Sur, Guerrero, México; H. H. Smith; date of capture not specified. Holotype.

Skin.—Faded. Facial stripes and color not distinguishable.
Skull.—Dental formula 2/2, 1/1, 2/2, 2/2; lower incisors without lobes on cutting edge; rostrofacial angle approximately 120°.

Remarks.—Andersen indicated that this specimen was a female. He examined a total of 12 specimens (including five skulls) from the Mexican states of Guerrero, Colima, Sinaloa, and Veracruz, but only the holotype was referred to by number.

Artibeus planirostris grenadensis Andersen

1906.   Ann. Mag. Nat. Hist., ser. 7, 18:420. December.

BMNH 96.11.8.6: adult male; in alcohol, skull removed; [Island of] Grenada, Lesser Antilles; T. J. Mann; date of capture not specified. Holotype.

Skin.—Faded somewhat.
Skull.—Molars 2/3.

Remarks.—Andersen examined "11 specimens and 8 skulls" from the island of Grenada but identified only the holotype by number.

Artibeus planirostris trinitatis Andersen

1906.   Ann. Mag. Nat. Hist., ser. 7, 18:420. December.

BMNH 97.6.7.1: adult female; skin and skull; St. Anns, Trinidad; Dr. Perry Rendall, no. 90, measured 90, 0, 13, 13; 23 February 1897. Holotype.

Skin.—Dorsal pelage gray; ventral pelage gray with white frosting.
Skull.—Molars 2/3.

Remarks.—Andersen examined a total of 13 specimens (including nine skulls) from the islands of Trinidad and Tobago but identified only the holotype by number.

Artibeus pumilio Thomas

1924.   Ann. Mag. Nat. Hist., ser. 9, 13:531. May.

BMNH 24.3.1.52: adult female; skin and skull; Tushemo, near Masisea, 1000 ft., Río Ucayali [in departament of Loreto], Perú; Latham Rutter, no. 228; 18 September 1923. Holotype.

Skin.—Facial stripes moderately developed, white, four; ears with pale border on part; tragus yellow.

Skull.—Molars 2/2; lower incisors distinctly bilobed, medial lobe approximately twice size of lateral lobe; compared with TCWC 12338, facial angle greater (approximately 136°) in holotype.

Remarks.—Thomas examined a single specimen. Tushemo is a place near Masisea, both on the Río Ucayali, a short distance up river from Pucallpa.

Artibeus (Dermanura) quadrivittatus Peters

1865.   Monatsb. Kön. preuss. Akad. Wiss. Berlin, p. 358.

RNH 13114: adult of undetermined sex; skin, skull removed; Surinam; H. H. Dieperink; date of capture not specified. Holotype or syntype.

Skin.—Faded. Four facial stripes; interfemoral membrane without fringe of hair on free edge, but some hair present on median ventral surface.
Skull.—Damaged. Molars 2/2; facial slope short, top of braincase slopes gradually from face beyond midpoint.

Remarks.—The name first appears as "*Dermanura quadrivittatum* nov. sp." in the title of the paper on p. 351.

Artibeus (Dermanura) Rosenbergi Thomas

1897.   Ann. Mag. Nat. Hist., ser. 6, 20:545. December.

BMNH 97.11.7.76: adult male; in alcohol, skull removed; Cachabí [in province of Esmeraldas], Ecuador; W. F. H. Rosenberg, no. 43; 17 December 1896. Holotype.

Skin.—Condition poor.
Skull.—M3 present on one side, absent on other; facial slope short, top of braincase slopes gradually to midpoint.

Remarks.—The locality appears to have been written "Cachuri" on the skull label, but Thomas reported the type locality as "Cachavi, N. Ecuador"; Cabrera (1958:88), as "Cachabí, provincia de Esmeraldas." The holotype probably came from the vicinity of the Río Cachabí, which joins the Río Bogotá at approximately 1°3'N, 78°50'W.

Artibeus turpis Andersen

1906.   Ann. Mag. Nat. Hist., ser. 7, 18:422. December.

BMNH 88.8.8.29: adult female; in alcohol, skull removed; Teapa, Tabasco, México; H. H. Smith; date of capture not specified. Holotype.

Skin.—Faded.
Skull.—Molars 2/2; lower incisors with cutting edge more serrate than lobed, similar to those of holotype for *A. nanus*; rostrofacial angle approximately 125°, similar to that of holotype for *A. nanus*.

Remarks.—Thomas examined three additional specimens, but none was identified by number. We found no differences to distinguish the skulls of holotypes for *A. turpis* and *A. nanus*.

Artibeus Watsoni Thomas

1901.   Ann. Mag. Nat. Hist., ser. 7, 7:542. June.

BMNH 0.7.11.19: adult male; skin and skull; Bogava [Bugaba], 250 m., Chiriquí, Panamá; H. J. Watson, no. 47; 24 October 1898. Holotype.

Skin.—Facial outline as in holotype for *Artibeus rosenbergi.*

Remarks.—Thomas examined three additional specimens, but none was identified by number.

Madataeus Lewisii Leach

1821.   Trans. Linn. Soc., London, 13:82.

BMNH *b* 4.b: juvenile, probably female; skin; Jamaica; W. Lewis; date of capture not specified. Holotype or syntype.

Skin.—Pelage apparently that of a juvenile, ventral pelage without frosting.
Skull.—Not located and presumed lost.

Remarks.—This specimen appears to have been old enough to fly at the time of its capture. Leach reported the collector as "D. Lewis."

Phyllostoma planirostre Spix

1823.   Simiarum et vespertilionum brasiliensium species novae . . . , p. 66, p. 36, fig. 1.

ZSM 66: adult of undetermined sex; partial skeleton; Salvador, Bahia, Brazil; Spix; date of capture not specified. Holotype or syntype.

Skin.—Not located and presumed lost.
Skull.—Molars 2/3.

Remarks.—ZSM 66 is a dry partial skeleton (skull, axial skeleton, pectoral girdle, and rib cage) probably that of a subadult inasmuch as the basioccipital and basisphenoid are not fused. The type locality is based on Spix' statement "habitat in suburbiis Bahiae" (now known as Salvador). A white label on the specimen bears the notation: "1903/9428 Bibia / aus Alkohol (Skelett) / Spix. coll. Type."

Enchisthenes

Artibeus Hartii Thomas

1892.   Ann. Mag. Nat. Hist., ser. 6, 10:409. November.

BMNH 92.9.7.8: juvenile male; in alcohol, skull removed; Botanic Gardens, Trinidad; J. H. Hart; date of capture not specified. Holotype.

Skin.—Condition poor.
Skull.—Damaged.

Remarks.—Thomas received a single specimen from Hart, superintendent of the Botanical Gardens in Trinidad.

Stenoderma

stenoderma rufa Desmarest

1820. Mammalogie ou description des especes de mammiferes, 1:117.

MNHN 934: adult of undetermined sex; skin; [here restricted to St. John, American Virgin Islands]; collector and date of capture unknown. Holotype.

Skin.—Faded. Dorsal pelage reddish, with three bands of color (dark, pale, dark); interfemoral membrane narrow, hairy, with fringe of hair; legs and forearms hairy; nose leaf no more than a dermal fold lateral to the nares—apparently an abnormality.

Skull.—Not located and presumed lost (see Genoways and Baker, 1972).

Remarks.—This specimen seems almost certainly to be the one on which É. Geoffroy St.-Hilaire (1818b) based the name "le Sténoderme roux," the basis for Desmarest's name *stenoderma rufa* (the generic name was not capitalized in the original description). Although Geoffroy St.-Hilaire originally reported this species as an inhabitant of Egypt, Desmarest stated that its place of origin was unknown. The label data now with the specimen indicate that it came from the Antilles, possibly from Puerto Rico ["Antilles (? Porto Rico)"], but we believe that this notation was added to the specimen label after Anthony (1918) reported fossil remains of *Stenoderma rufum* from Puerto Rico. Hall and Bee (1960) reported three specimens of S. *rufum* from St. John, American Virgin Islands, and Hall and Tamsitt (1968) described S. *rufum darioi* from Puerto Rico and applied the name S. *r. rufum* to specimens from St. John and St. Thomas. Because the holotype of S. *rufum* agrees in color with the specimens from St. John (see Jones *et al.*, 1971) and, therefore, is unlikely to have come from Puerto Rico, we restrict the type locality of the nominate subspecies to St. John. The holotype for S. *rufum* was not included in Rode's (1941) catalogue of types in the Paris museum.

Pygoderma

Phyllostoma bilabiatum Wagner

1843. Arch. Naturgesch., 9(1):366.

ZSM 114: adult female; skin, skull not removed; Ypanema [Ipanema, in state of São Paulo], Brazil; Johann Natterer; 22 December 1819. Holotype.

Skin.—Faded somewhat. Dorsal pelage pale brown with slight reddish cast, individual hairs with basal band brownish, median band whitish, distal band pale brown with reddish cast; ventral pelage much paler, individual hairs with basal band pale brown, distal band pale gray white; white shoulder spot; interfemoral membrane hairy on dorsal and ventral surfaces, free edge with fringe of hair.

Skull.—In skin; apparently entire.

Remarks.—"N. 64 / Ypanema / 22 Decbr. 19 / Foemina" is inscribed on the original label; an old green label bears the inscription "*Phyllostoma bilabiatum* Natt. / foemina / 1844 / Ypanema." Penciled on the green label, and possibly in Peter's handwriting, is the note "*Pygoderma* Ptrs."

Stenoderma (Pygoderma) microdon Peters

1863.   Monatsb. Kön. preuss. Akad. Wiss. Berlin, p. 83.

ZMB 2713: adult male; in alcohol, skull removed; Surinam; collector and date of capture unknown. Syntype.

Remarks.—The Berlin museum catalog identifies 2712 (also in alcohol) as "typus," but this specimen could not be located and is presumed lost. Notwithstanding, inasmuch as Peters made no distinction between the two specimens upon which he based the name *Stenoderma (Pygoderma) microdon,* the two have equal status as syntypes. Both came from Surinam.

Ametrida

Ametrida centurio Gray

1847.   Proc. Zool Soc., London, p. 15.

BMNH 1957.a.: adult female; in alcohol, skull removed; Pará [now Belem], Pará, Brazil; J. P. G. Smith; date of capture not specified. Holotype.

Remarks.—This specimen must be the same as that listed in Dobson (1878:531) as "*a.♂*ad., al." from Pará, which was obtained from J. P. G. Smith. The holotype, however, is a female; Gray did not specify the sex of the single specimen upon which he based the name *Ametrida centurio.*

Sphaeronycteris

Sphaeronycteris toxophyllum Peters

1882.   Sitzungsb. preuss. Akad. Wiss., p. 989, pl. 16.

ZMB 598: adult female; in alcohol, skull removed; South America [Pebas, Loreto, Perú, by restriction (Cabrera, 1958:92)]; collector and date of capture unknown. Holotype.

Skin.—Dorsal pelage pale reddish, with basal band pale reddish and median band white; venter buff, pelage pale to whitish basally, gular region whitish; nose leaf small but distinct, broadly rounded on top, almost 2 millimeters long measured from back; white spot at junction of humerus and scapula.

Skull.—Not located and presumed lost.

Remarks.—Peters examined a single specimen, which he reported to be number 5984. Without a doubt, this is the one now numbered 598.

Centurio

Centurio senex Gray

1842.   Ann. Mag. Nat. Hist., 10:259. 31 December.

BMNH (not numbered): adult female; in alcohol, skull removed; Amboyna, according to Gray; [Realejo, Chinandega, Nicaragua, by restriction (Goodwin, 1946:327)]; Sir Edward Belcher; date of capture not specified. Holotype.

Skull.—Slightly damaged.

Remarks.—This specimen was collected on the voyage of H.M.S. Sulphur, 1836-1842, under the command of Captain Sir Edward Belcher.

Phyllonycterinae

Brachyphylla

Brachyphylla cavernarum Gray

1834.   Proc. Zool. Soc., London, p. 123. 12 March.

BMNH (not numbered): adult male; in alcohol; St. Vincent, Windward Islands, Lesser Antilles; Lansdown Guilding; date of capture not specified. Syntype.

Skin.—Pelage slightly faded.

Remarks.—Gray originally had two specimens (one male and one female) from St. Vincent, both from the collection of the late Rev. Lansdown Guilding, a corresponding member of the Zoological Society, who apparently collected the specimens. A third specimen (unnumbered) now in the British Museum and long labeled as "type" of *Brachyphylla cavernarum* is a fluid-preserved subadult male of the species now known as *Brachyphylla nana*; it is from Cuba and, although listed by Gray (1843:20), has no type status.

Desmodontinae

Desmodus

Desmodus D'Orbignyi Waterhouse

1838.   The zoology of the voyage of H.M.S. Beagle . . . , 1(Mammalia):1, pls. 1, 35, figs. 1a-g (skull).

BMNH (not numbered): adult female; skin; Coquimbo, Chile; Charles Darwin; date of capture not specified. Holotype.

D[esmodus]. murinus Wagner

1840.   Die Säugthiere . . . von Schreber, Supplementband, 1:377.

ZSM 57: adult male; skin, skull not removed; "Brazil"; collector uncertain (probably Johann Natterer); date of capture not specified.

ZSM 59: adult female; skin, skull not removed; Cuyabá [Cuiabá, Mato Grosso, Brazil]; Johann Natterer; 26 May 1824.

ZSM 131: juvenile male; skin; Cuyabá [Cuiabá, Mato Grosso, Brazil]; Johann Natterer; date of capture not specified.

Skin.—All three faded.

Skull.—Those for ZSM 57 and 59 apparently entire; that for 131 not located and presumed lost.

Remarks.—All three specimens were labeled "typus," probably subsequent to World War II, but it is unlikely that any of these is one of the two Mexican specimens referred to by Wagner (p. 379) as having come from the Würzburg Museum. Specimen labels for ZSM 57 indicate that it was taken by Johann

Natterer and sent on exchange in 1840 to the Berlin museum; ZSM 59, and probably 131, are the two specimens sent by Joseph Natterer to Wagner in 1843. None is a type specimen for the name *Desmodus murinus*, and we are reasonably certain that none now exists in the Berlin museum.

Edostoma cinerea d'Orbigny

1835.   Voyage dans l'Amérique Méridionale . . . années 1826 . . . 1833, 9(Atlas Zoologique), pl. 8.

MNHN 958D: adult of undetermined sex; skin and skull; Santa Fe de Bogotá [Colombia?]; collector and date of capture unknown. Syntype?

MNHN 958G: adult of undetermined sex; skin, skull not removed; locality, collector, and date of capture unknown. Syntype?

Remarks.—MNHN 958D was listed by Rode (1941:97) as the holotype (number 227) of *Edostoma cinerea* and thought to be an example of *Molossus rufus* É. Geoffroy St.-Hilaire. Notwithstanding, the specimen is one of *Desmodus*, but probably it is not a type of *E. cinerea* inasmuch as d'Orbigny's specimen(s) came from Bolivia. The specimen (958G, type number 227a) listed by Rode as a paratype of *E. cinerea* is from an unknown locality although "Sta. Fe de Bogota?" appears on the label. The Paris museum catalogues, all written some years after these two specimens arrived at the museum, shed no light on the origin or status of either. Both skins appear to have been prepared from specimens originally preserved in spirits. Type 227 also is numbered A.342 and 1018. Date of publication for pl. 8 of Atlas Zoologique was taken from Sherborn and Griffin (1934).

Diaemus

Desmodus Youngii Jentink

1893.   Notes Leyden Mus., 15:282. October.

RNH 12088: adult male; in alcohol, skull removed; upper Canje Creek, "Berdice" [a tributary of the Berbice River], British Guiana [Guyana]; Dr. C. G. Young; 1893. Holotype.

Skull.—Cheek teeth 2/3.

Remarks.—A single specimen was examined.

Diaemus youngi cypselinus Thomas

1928.   Ann. Mag. Nat. Hist., ser. 10, 2:288. September.

BMNH 28.7.21.64: adult female; skin and skull; Pebes [Pevas], 300 ft., Loreto, Perú; R. W. Hendee (Godman-Thomas Expedition), measured 99, 0, 19.5, 18.5, forearm 56; 28 January 1928. Holotype.

Remarks.—The notation "in house in evening" is inscribed on the specimen label. Thomas examined a single specimen. Pevas is situated on the north bank of the Río Marañon at 3° 10′S, 71° 49′W.

Diphylla

Diphylla centralis Thomas

1903.   Ann. Mag. Nat. Hist., ser. 7, 11:378. April.

BMNH 3.3.3.3: adult male; skin and skull; Boquete, 4500 ft., Chiriquí, Panamá; H. J. Watson, no. 62; 4 March 1902. Holotype.

Remarks.—The specimen label is inscribed with the notation "caught in low bush." The name *Diphylla centralis* was based on a single specimen.

NATALIDAE

Natalus

Natalus stramineus Gray

1838.   Mag. Zool. Bot., London, 2:496.

BMNH (not numbered): adult male; in alcohol, skull removed; Brazil [Lagoa Santa, Minas Gerais, by restriction (Cabrera, 1958:95)]; collector and date of capture unknown. Syntype.

Skin.—Condition fair.

Remarks.—The name *Natalus stramineus* was based on material in the British Museum, but Gray's statement (p. 496) "inhabits ——?" suggests that the origin of his specimen(s) was unknown; Later, Gray (1843:28) listed two British Museum specimens of *N. stramineus,* one in fluid from "South America" (now labeled Brazil, the only identifiable one remaining) and another from "N. America, St. Blas," apparently a skin. The fluid-preserved specimen with skull removed is here considered to be Gray's (1843:28) specimen "*a*" and a syntype for *N. stramineus.* Goodwin (1959:2-4) disagreed with Cabrera and stated that the type locality was "probably Antigua," Lesser Antilles.

Phodotes tumidirostris continentis Thomas

1911.   Ann. Mag. Nat. Hist., ser. 8, 7:513. May.

BMNH 11.5.25.13: adult male; skin and skull; San Esteban, Carabobo, Venezuela; S. M. Klages, no. 77, measured 43, 48, 6.5, 14; 24 December 1910. Holotype.

Remarks.—Thomas examined three additional specimens, all females apparently collected by S. M. Klages, field numbers for which were given as 50, 78, and 124.

THYROPTERIDAE

Thyroptera

Hyonycteris discifera Lichtenstein and Peters

1855.   Monatsb. Kön. preuss. Akad. Wiss. Berlin, 1854:335.

ZMB 59067: adult male; in alcohol; Puerto Cabello [in state of Carabobo], Venezuela; von Appun; date of capture not specified. Lectotype (Wilson, 1976).

Remarks.—We located two of the three specimens (all males) that von Appun sent to the Berlin museum. See Wilson (1976) for a discussion of these specimens.

Th[*yroptera*]. *bicolor* Cantraine

1845.   Bull. Acad. Roy. Sci. Belles-lettres, Bruxelles, 12:492.

RNH 17551: adult female; in alcohol, skull removed; Surinam; H. H. Dieperink; date of capture not specified. Holotype.

Remarks.—Although Cantraine (p. 495) stated that the single specimen examined was a male, the one numbered RNH 17551 and identified as the type for *T. bicolor* is a female. This specimen was listed in Jentink (1888) as "*Th. tricolor,* no. *b.*" H. H. Dieperink was a military physician in Surinam.

Vespertilionidae

Myotis

Myotis chiloensis alter Miller and G. M. Allen

1928.   Bull. U.S. Nat. Mus., 144:194.

BMNH 0.6.29.23: adult female; in alcohol, skull removed; Palmeira, Paraná, Brazil; G. Grillo, from Genoa museum; date of capture not specified. Holotype.

Skin.—Condition fair, some loss of hair.

Remarks.— The name *Myotis oxyotis alter* is written on the label. In addition to the holotype, there are 20 paratypes, four from Brazil and 16 from Argentina (see Miller and Allen, 1928:195-196, for museum numbers and measurements).

Myotis Dinellii Thomas

1902.   Ann. Mag. Nat. Hist., ser. 7, 10:493. December.

BMNH 0.7.9.4: adult female; skin and skull; Tucumán, 456 m., Argentina; L. Dinelli, no. 17, measured 44.5, 41.5, 8, 14; 9 April 1899. Holotype.

Skin.—Dorsal pelage reddish, hairs with basal one-half dark gray, distal one-half reddish; ventral pelage pale buff (basal half of hair dark gray); chiropatagium with fur on ventral surface from elbow to body.

Remarks.—"Caught with net in bedroom" is noted on the skin label. In addition to the holotype, Thomas examined four specimens, all collected from the type locality; none was identified by number. The date of capture was given as 4 April in the original description.

Myotis lucifugus fortidens Miller and G. M. Allen

1928.   Bull. U.S. Nat. Mus., 144:54.

BMNH 88.8.18: adult female; in alcohol, skull removed; Teapa, Tabasco, México; H. H. Smith; 5 January 1888. Holotype.

Skin.—Condition fair, some loss of hair.

Remarks.—Hall and Dalquest (1950:585) assigned the single paratype, a female (USNM 21083/36121) from Fort Hancock, El Paso County, Texas, to *Myotis lucifugus carissima* Thomas, 1904; later (Hall and Kelson, 1959:167), referred to *Myotis occultus* Hollister, 1909.

Myotis simus Thomas

1901.   Ann. Mag. Nat. Hist., ser. 7, 7:541. June.

BMNH 81.5.12.2: adult female; in alcohol, skull removed; Sarayacu [6°44′S, 75°6′E, in department of Loreto], Perú; W. Davis; 1876. Holotype.

Skin.—Faded somewhat, some loss of hair.
Skull.—Sagittal crest relatively well developed; occiput elevated; rostrum broad.

Remarks.—Sarayacu is a small settlement located approximately 11 kilometers west of the Río Ucayali. Thomas examined a single specimen and indicated that it was a female.

Myotis surinamensis Husson [substitute name for *Vespertilio ferrugineus* Temminck, 1840]

1962.   Zool. Verhand., Leiden, 58:218, figs. 21f, 35; pl. 24.

RNH 17363: adult male; skin, skull removed; Surinam, H. H. Dieperink; date of capture not specified. Lectotype.

Remarks.—See account of *Vespertilio ferrugineus* Temminck for additional information.

Myotis thysanodes aztecus Miller and G. M. Allen

1928.   Bull. U.S. Nat. Mus., 144:128.

BMNH 58.6.2.3: adult of undetermined sex; skin and skull; San Antonio, Oaxaca, México; collector and date of capture unknown. Holotype.

Remarks.—Two notations inscribed on the skin label by someone other than the collector are: "*Myotis aztecus* Tomes, 1860" (in apparent reference to the specimen having belonged to Tomes at one time; the meaning of 1860 is unknown to us) and "*Vespertilio albescens* Geoff= V. *leucogaster* Wied= V. *nubilus* Wagner." There are five paratypes, all of undetermined sex, from Hacienda de Cinco Señores, Oaxaca, México (BMNH 58.6.2.4; 7.1.1.523-526).

Vespertilio arsinoe Temminck

1840.   Monographies de Mammalogie, 2:247.

RNH 17635: adult female; skin, skull removed; Surinam; collector and date of capture unknown. Holotype.

Skin.—Condition poor, faded somewhat. Dorsal pelage reddish brown, hair darkest at base, becoming paler distally, tipped with pale buff; venter pale orange, hairs with basal two-thirds reddish brown; uropatagium with hair on proximal portion above and below;

calcar apparently without keel; compared with TCWC 12682, 12702, 12703, 12710, 12713, 12715, 12727, 12737, holotype of *V. arsinoe* and TCWC 12702 appear to be referable to the same species.

Skull.—Condition poor. Rostrum relatively short and broad; upper premolars crowded, last upper premolar overlaps second laterally; skull of holotype resembles somewhat that of TCWC 12682, but occiput not so elevated, upper premolars more crowded, and rostrum shorter; skull of holotype most closely resembles that of TCWC 12702 in elevation of occiput and in upper premolars, but frontal portion more inflated.

Remarks.—Sex cannot be determined with certainty from the mounted skin, but female is written on the wooden base to which the specimen is attached, and Temminck stated (p. 248) that the single specimen received by the Pays-Bas museum was an old female.

Vespertilio carbonarius Wagner

Possibly a published name.

ZSM 124: adult of undetermined sex; skin; Brazil; Brandt; date of capture not specified. Type?

Skin.—May be slightly faded (membranes reddish brown). Dorsum dark reddish brown, basal three-fourths of hair black; venter paler, basal three-fourths of hair black, distal one-quarter pale reddish brown to dull buff (paler than in TCWC 12715); except for possible fading, pelage of *V. carbonarius* is much like that of TCWC 12715; ears, tragus, feet, and calcar much like those of TCWC 12715 (ZSM 124 and TCWC 12715 appear to belong to the same species). Type also compared with TCWC 12682, 12702, 12703, 12713, 12727, and 12737.

Skull.—Not located and presumed lost.

Remarks.—An old green label bears the notation "*Vespertilio carbonarius* Wagn. / 1843 / Brandt / Brasil." Although the name was not included in Wagner (1843, 1845, 1847, 1855), it might have been published elsewhere, inasmuch as we did not examine all of Wagner's works. Brandt was a dealer in natural history objects.

Vespertilio Chiloensis Waterhouse

1838.   The zoology of the voyage of H.M.S. Beagle . . . , 1(Mammalia):5, pl. 3, 35 (figs. 3a-c).

Remarks.—Specimen presumed lost. Miller and Allen (1928) were unable to locate it and so were we. A neotype (adult female, Field Museum of Natural History no. 24029 for this species) was designated by LaVal (1973:43).

Vespertilio ferrugineus Temminck [replaced by the substitute name *Myotis surinamensis* Husson, 1962]

1840.   Monographies de Mammalogie, 2:239, pl. 59, fig. 2.

RNH 17363: adult male; skin, skull removed; Surinam; H. H. Dieperink; date of capture not specified. Lectotype (Husson, 1962:218).

Skin.—Condition poor; faded. Hair on dorsum with basal two-thirds dark brown, distal one-third buff; hair on venter with basal two-thirds dark brown, distal one-third cream; tragus moderately long, tapering to point.

Skull.—Damaged. The size and general appearance of the skull is like that of *Myotis dasycneme.*

RNH 17364: adult male; skeleton; Amerique meridionale; collector and date of capture not specified. Paralectotype (see Avila-Pires, 1965:8).

Remarks.—Husson (1962) gave the sex of RNH 17363 ("no. *a*" in Jentink, 1888:187) as female. RNH 17364 is Jentink's (1887:281) "no. *a*," and one of the original syntypes on which Temminck based the name *V. ferrugineus.* It is doubtful that either of these specimens came from South America (see LaVal, 1973:46); furthermore, both resemble *M. dasycneme* (according to James S. Findley, personal communication), a species occurring in northern Europe. Whatever the case, the name *V. ferrugineus* is not available because it is a junior primary homonym of *V. ferrugineus* C. L. Brehm, 1827, and Husson (1962) renamed this species *Myotis surinamensis.*

Vesp [*ertilio*]. *leucogaster* Schinz

1821.   Das Thierreich . . . , 1(Säugethiere und Vögel):180.

RNH 17622: subadult female; skin, skull not removed; Mucurí [in state of Bahia, 18° 43′S, 39° 46′W], Brazil; Maximilian, Prinz zu Wied-Neuwied; date of capture not specified. Paralectotype (see Avila-Pires, 1965:8).

Skin.—Faded somewhat. Individual hairs on dorsum blackish at base, becoming reddish distally, with tips dirty yellow; hair on venter with basal two-thirds blackish to dull dark reddish brown, distal one-third dull buff; pelage, ear, and tragus similar to those of TCWC 12682.

Skull.—All permanent teeth present; small space between second and third upper premolar; first and second upper premolars appressed, as in TCWC 12682.

Remarks.—Jentink (1887:283) first listed RNH 17622 as "no. *c*"; later (1888:192), as "no. *b*: *Vespertilio albescens.*"

Authors have long attributed the name *Vespertilio leucogaster* to Wied-Neuwied (1826:271), probably because Schinz credited the name to "P. Max" (Maximilian, Prinz zu Wied-Neuwied), but Wied-Neuwied (*loc. cit.*) cited Schinz (1821:180) and "Abbildungen zur Naturgeschichte Brasilien's" (Wied-Neuwied, 1822-1831, pl. 21 and text on facing page). So far as we can determine, the first description for *V. leucogaster* is that by Schinz (1821:180)—technically, Schinz is the author of all names that he credited to "P. Max." (however, Wied-Neuwied is the author of those names for which Schinz cited "Neuwied").

The specimen (RNH 17622) now labeled as cotype for this name is that of a *Myotis*, and, although Wied's (1826:274) description of its habits and teeth and his placement of the account between those of *V. calcaratus* and *V. naso* suggest an emballonurid, the description matches both that of Schinz (1821) and the specimen numbered RNH 17622. The specimen was prepared in such a way (mounted skin with skull in place) that Schinz and Wied-Neuwied would have had difficulty examining the cheek teeth, so that these easily could have been miscounted.

Cabrera (1958) listed the type locality as Río Moucourí; however, the origin of his information is unknown to us. Inasmuch as Wied is known to have been collecting along the Brazilian coast in the states of Bahia and Espírito Santo, we believe the location of the type locality noted on the specimen label to be correct. The lectotype is no. 385, American Museum of Natural History.

Vespertilio levis I. Geoffroy St.-Hilaire

1824.   Ann. Sci. Nat. Paris, 3:444.

MNHN 864: adult of undetermined sex; skin and partial skull; Brazil; Auguste de St.-Hilaire; date of capture not specified. Holotype or syntype.

Skin.—Ears long, relatively narrow; tragus long, narrow, and pointed.

Remarks.—This specimen was taken somewhere in southern Brazil, as were all of those collected by Auguste de St.-Hilaire (records of the Muséum National d'Histoire Naturelle). The type number is 203.

V[*espertilio*]. *nubilus* Wagner

1855.   Die Säugthiere . . . von Schreber, Supplementband, 5:752.

ZSM 121: subadult of undetermined sex; skin, skull not removed; Brazil; collector and date of capture unknown. Holotype or syntype.

Skin.—Faded somewhat. Hair on dorsum reddish brown, not banded although tips paler and somewhat shiny; hair on venter paler than above, basal three-fourths or four-fifths dark brown, tips (distal one-fourth or one-fifth) buff to buffy white; chiropatagium with long hair on ventral surface proximal to elbow; except for fading, membranes, ear, tragus, calcar, and foot much like those of TCWC 12703.

Skull.—Apparently entire; mouth open, teeth visible. Upper premolars as in TCWC 12703.

Remarks.—An old green label bears the notation "*Vespertilio nubilus* Wagn / (*V. albescens* Tem. *nec* Geoffr.) / 1842 / Bras." Epiphyses on wing bones are almost closed.

Vespertilio parvulus Temminck

1840.   Monographies de Mammalogie, 2:246.

RNH 17621: adult of undetermined sex; skin, skull removed; Brazil; "Natterer (from Wied)"; date of capture not specified. Lectotype (Husson, 1962:210).

Skin.—Hair on dorsum dark reddish brown, with basal half blackish; uropatagium with hair on dorsal surface extending short distance beyond knees.
Skull.—Incomplete.

Remarks.—Jentink (1887:283) listed this specimen as *Vespertilio nigricans*, "no. *a*" and stated that it came from the Natterer collections by way of Maximilian, Prinz zu Wied-Neuwied ["Natterer (from Wied)" is noted on the label]. Temminck (pp. 246-247) stated that the specimens upon which the description of *Vespertilio parvulus* was based came from the "Musée de Vienne et Pays-Bas" and, according to "Mr. Natterer de Vienne," were captured in Brazil. Husson (1962:210) stated that RNH 17621 was collected by J[ohann]. C. Natterer.

Vespertilio (Leuconoë) pilosus Peters

1869.   Monatsb. Kön. preuss. Akad. Wiss. Berlin, p. 403.

MNHN (not numbered): adult female; in alcohol, skull removed; "Montevideo"; Lassaux; date of capture not specified. Holotype.

Remarks.—Although Miller and Allen (1928:209) reported that the type for *V. pilosus* could not be found, Rode (1941:92) listed a specimen (type no. 209, the same as reported here) purported to be the holotype for this name. Rode stated, as did Peters (p. 405), that the specimen was received from Lassaux. It appears likely that the specimen was purchased from Lassaux and that the locality data are erroneous. Wherever it came from, it is almost certain that it was not South or North America (see Miller and Allen, 1928:209). According to Rode (1941), the skull had not been removed at the time he examined it.

Vespertilio polythrix I. Geoffroy St.-Hilaire

1824.   Ann. Sci. Nat. Paris, 3:443.

MNHN 842: adult of undetermined sex; skin, skull not removed; Brazil; Auguste de St.-Hilaire; date of capture not specified. Syntype.

MNHN 843: adult male; skin, skull not removed; Brazil; Auguste de St.-Hilaire; date of capture not specified. Syntype.

Remarks.—The specimens on which this name was based apparently came from two general localities, Rio Grande do Sol and Minas Gerais (Miller and Allen, 1928:197). Only two specimens (842 and 843 are identified as types, but three additional specimens, all subadults (844-846), are labeled "*V. polythrix* Is. Geoffr." and appear to be examples of the original series.

Vespertilio splendidus Wagner

1845.   Arch. Naturgesch., 11(1):148.

ZSM 142: adult of undetermined sex; skin; St. Thomas [American Virgin Islands]; Schimper; date of capture not specified. Holotype or syntype.

Skin.—May be somewhat faded. Dorsum medium reddish brown (more reddish than brown), hair slightly darker basally; venter paler than dorsum, hair with basal three-fourths moderately dark reddish brown; uropatagium with hair on proximal one-fourth to one-third of dorsal surface as in *Myotis nigricans*; feet slightly larger than those of TCWC 12703; tragus relatively broad from base to midportion, tapers rather abruptly to point; calcar with moderately elongate keel.

Remarks.—The name *V. splendidus* seems to have been overlooked by most subsequent workers. Cabrera (1958:95) put this name in synonymy with *Natalus stramineus*; notwithstanding, the type is a specimen of *Myotis*, possibly *nigricans*. The oldest of two green labels bears the notation "*Vespertilio splendidus* Wagn. / 1844 / Ins. St. Thomas"; according to the more recent label, this specimen was taken by Schimper.

Pipistrellus

Vespertilio isidori d'Orbigny and Gervais

1847.   Voyage dans l'Amérique Méridionale . . . années 1826 . . . 1833, 4(2):16.

MNHN 865: adult of undetermined sex; skin and skull; Brazil; Davenne; date of capture not specified. Not a type.

Skin.—Faded. Hair with four bands of pigment (base to tip, dull gray, pale yellow, pale brown, pale yellow); ventral pelage apparently dark basally and pale (perhaps yellow) distally.

Skull.—First upper incisor with secondary cusp; I2 slightly smaller than I1; first upper premolar small, in toothrow, with simple cusp; i1 broken on both sides; i2 either bilobed or indistinctly trilobed; i3 appears larger than i2, obviously trilobed; first lower premolar about two-thirds size of second.

Remarks.—Although Rode (1941:91) listed this specimen as the holotype for *Vespertilio isidori*, he justifiably questioned its status as such, probably because he was confused (as are we) by the Paris museum's catalogue references to the number 865. A label with this specimen identifies it as "V. Isidori, d'Orb. / M. Davenne Brésil / 865"; presumably, 865 is a "cadre" number because Davenne's specimen appears in the catalogue as number "544, V. Isidorii, Brésil, Davenne 1842, 865 (cadre)." [Cadre numbers, of which there were several series, were given to specimens mounted for display in the public gallery; these numbers, written on the wings of bats, corresponded to numbers in printed catalogues intended for use by visitors to the museum.] Also listed in the Paris museum catalogue is a specimen of "*Vesperugo cinnamomeus*, Amérique Mérid, Castelnau, 1844, 865 (cadre)"; obviously, there were at least two bats inscribed with a cadre number of 865. If one assumes that the specimen labeled "holotype?" by Rode is the one obtained from Davenne and purported to be from Brazil, one would have to conclude that it would not be a type for the name *V. isidori* because Gervais (1855) stated that d'Orbigny and Gervais based this name on one or more specimens from Corrientes, Argentina. But more importantly, it is unlikely that the specimen at hand even came from South America inasmuch as it is referable to the genus *Pipistrellus*, a taxon not known to occur on that continent.

Vespertilio lacteus Temminck

1840.   Monographies de Mammalogie, 2:245.

RNH 17624: adult of undetermined sex; skin, skull removed; "?Amérique méridionale?"; collector and date of capture unknown. Syntype.

Skin.—Condition poor, apparently prepared from fluid-preserved specimen; no reliable measurements can be taken; faded. Hair above and below with dark brown basal band and dirty white (or very pale yellow) distal band.

Skull.—Damaged, few measurements possible.

Remarks.—Jentink (1887:277; 1888:178) listed this specimen as *h* and *a*, respectively, of *Vesperugo pipistrellus*. *Vespertilio lacteus* was not listed by Cabrera (1958), and, although the specimen is in such poor condition that it cannot be identified with certainty, it appears not to have come from the New World. A notation, "=???*Pipistrellus pipistrellus bactrianus* Satunin," accompanies the specimen. Temminck examined two specimens from an unspecified locality, which he thought probably to be "Amérique méridionale."

Eptesicus

Eptesicus argentinus Thomas

1920.   Ann. Mag. Nat. Hist., ser. 9, 5:365. April.

BMNH 98.3.4.6: adult female; skin and skull; Goya, on the [Río] Paraná, 600 ft., Corrientes, Argentina; R. Perrens, no. 18, measured 67, 44, 11, 18; 16 December 1895. Holotype.

Skin.—Dorsum pale reddish brown, hairs darker brown basally; overall appearance rather like that of *Eptesicus fuscus* except membranes not so dark, hair shorter, and tragus longer and slightly more pointed.
Skull.—Zygomata broken, otherwise condition good; Thomas reported the zygomatic breadth to be 12.5. Sagittal crest present, helmet well developed.

Remarks.—This specimen appears to have been lactating. The notation "common, killed in roof of house" was written on the skin label. In addition to the holotype, Thomas examined seven specimens, none of which was identified by number.

Eptesicus chiriquinus Thomas

1920.   Ann. Mag. Nat. Hist., ser. 9, 5:362. April.

BMNH 3.3.3.1: adult male; skin and skull; Boquete, 4000 ft., Chiriquí, Panamá; H. J. Watson, no. 90, measured 70, 50, 10, 14; 6 April 1902. Holotype.

Skin.—General appearance like that of *Eptesicus montosus* Thomas.
Skull.—Like that of *Eptesicus montosus.*

Remarks.—Thomas apparently examined a single specimen.

Eptesicus fidelis Thomas

1920.   Ann. Mag. Nat. Hist., ser. 9, 5:366. April.

BMNH 1.2.4.1: adult male; in alcohol, skull removed; Esperanza, Santa Fe, Argentina; E. Linder; date of capture not specified. Holotype.

Skin.—Condition fair; some loss of hair.
Skull.—Sagittal crest and helmet lacking.

Remarks.—Thomas examined a single specimen, for which the collector's name was given as E. Lindner.

Eptesicus fuscus pelliceus Thomas

1920.   Ann. Mag. Nat. Hist., ser. 9, 5:361. April.

BMNH 98.7.1.28: adult female; skin and skull; Culata, 4000 m., Mérida, Venezuela; Briceño, no. 64; 20 June 1897. Holotype.

Skin.—Dorsum medium reddish brown.
Skull.—Damaged to the extent that few measurements can be taken.

Remarks.—According to Thomas, the single specimen was taken by Sr. Briceño on 20 June 1897 at La Culata, a place situated on the "heights near [the city of] Mérida."

Eptesicus inca Thomas

1920.   Ann. Mag. Nat. Hist., ser. 9, 5:363. April.

BMNH 94.8.6.1: adult male; in alcohol, skull removed; Chanchamayo, Cuzco, Perú; J. Kalinowski; date of capture not specified. Holotype.

Skin.—Condition good but not measured; forearm 46 according to Thomas.
Skull.—Zygomata broken. Sagittal crest present, moderately developed.

Remarks.—Thomas examined a single specimen.

Eptesicus montosus Thomas

1920.   Ann. Mag. Nat. Hist., ser. 9, 5:363. April.

BMNH 2.1.1.1: adult male; skin and skull; Choro [66° W, 16° S], 3600 m., Cochabamba, Bolivia; Perry O. Simons, no. 1433, measured 55, 43, 10, 18, tragus 9; 8 May 1901. Holotype.

Skin.—Dorsum black with dark reddish cast; hair with basal four-fifths black, tips dark red; venter obscure buff or orange; hair with basal three-fourths black; ears and membranes black.
Skull.—Zygomata broken; zygomatic breadth 10.3 according to Thomas. Skull in general similar to that of *Eptesicus inca* Thomas.

Remarks.—Thomas apparently examined a single specimen. "Flying in evening" is noted on the skin label. Choro, in the highlands of Bolivia, is situated on the upper waters of the Río Mamoré, north of the town of Cochabamba.

Eptesicus punicus Thomas

1920.   Ann. Mag. Nat. Hist., ser. 9, 5:364. April.

BMNH 99.8.1.1: adult male; skin and skull; Puná, 10 m., Isla de Puná, Golfo Guayaquil, Ecuador; Perry O. Simons, no. 1, measured 42, 35, 7, 13; 1 November 1899. Holotype.

Skin.—Dorsum reddish brown, hair with basal one-half or two-thirds dark brown, venter buff to pale gray buff, hair with basal two-thirds dark brown.

Remarks.—"Shot in road" is noted on the skin label. Thomas recorded a forearm of 37 for the single paratype but gave no other details.

Scotophilus Cubensis Gray

1839.   Ann. Nat. Hist., 4:7. September.

BMNH 103*a*: adult female; in alcohol; Cuba; W. S. MacLeay; date of capture not specified. Holotype.

Skin.—Condition fair, some loss of hair.
Skull.—Firm, apparently entire.

Remarks.—This specimen is identified in the British Museum collection by the name *Eptesicus cubensis.* Gray examined a single specimen.

Scotophilus MacLeayii Gray

1843.   List of the specimens of Mammalia in the collection of the British Museum, p. 30. Apparently a *nomen nudum.*

BMNH 104a: adult male; in alcohol; Cuba; W. S. MacLeay; date of capture not specified.

Skin.—Condition fair, some loss of hair.

Remarks.—We found no published reference to this name other than the one above. The specimen is identified in the British Museum as a type for the name *Eptesicus macleayii.*

V[espertilio]. arctoideus Wagner

1855.   Die Säugthiere . . . von Schreber, Supplementband, 5:758.

ZSM 144: adult of undetermined sex; skin, skull not removed; Brazil; Brandt; date of capture not specified. Holotype.

Skin.—Dorsum dark reddish brown, hair with basal three-fourths dark brown; tragus moderately long, gently rounded.
Skull.—Posterior half of braincase missing.

Remarks.—An old green label is inscribed "*Vespertilio arctoideus* Wagn. / (?*V. polythrix* Is Geoffr.) / Bras." Wagner examined a single specimen, which was obtained from a dealer (Brandt) in natural history objects.

Vespertilio fuscus peninsulae Thomas

1898.   Ann. Mag. Nat. Hist., ser. 7, 1:43. January.

BMNH 95.34.10.14: adult male; skin and skull; Sierra Laguna, Baja California [Sur], México; D. Coolidge, no. 385; 7 July 1896. Holotype.

Remarks.—Thomas examined three specimens in addition to the holotype, but these were not identified by number. The type locality is located on the southern tip of Baja California Sur.

Vespertilio Hilarii I. Geoffroy St.-Hilaire

1824.   Ann. Sci. Nat. Paris, 3:441.

ZMB 3912: adult of undetermined sex; skin, skull not removed; Goyar, Missiones, Brazil; Auguste de St.-Hilaire; date of capture not specified. Syntype.

Skin.—Faded. Dorsum reddish brown, hair with basal two-thirds dull reddish brown, distal one-third paler reddish brown; hair on venter with basal two-thirds darker brown than on dorsum, distal one-third buff.

Remarks.—The Berlin museum obtained this specimen through an exchange with the Paris museum. None of the other original specimens taken by Auguste de St.-Hilaire could be found at the Paris museum, and, so far as we can determine, this is the only remaining one. I. Geoffroy St.-Hilaire (p. 442) stated that the specimens came from "Goyar . . . province des Missions."

Vesperugo (Vesperus) dorianus Dobson

1885.   Ann. Mus. Civ. Stor. Nat. Genova, 22:17.

BMNH 86.11.3.13: adult female; in alcohol, San Ignazio, Missiones, Argentina; Bore; November 1883. Syntype.

Skin.—Some loss of hair.
Skull.—Apparently entire.

Remarks.—This specimen was obtained from the Genova museum.

Vesperus melanopterus Jentink

1904.   Notes Leyden Mus., 24:176.

RNH 12092: adult female; in alcohol, skull removed; Paramaribo, Surinam; collector unknown; September 1903. Holotype.

Skin.—Dorsal pelage blackish to dark reddish brown basally, becoming reddish distally; ventral pelage reddish brown basally, becoming very pale (almost white) distally; tragus long and pointed.
Skull.—Damaged. First upper incisor much larger than second; three lower incisors trilobed, overlapping.

Remarks.—Jentink examined a single specimen. Dr. M. Greshoff, identified on the specimen label, was Director of the Colonial Museum at Haarlem; he sent this and several other bats to Jentink for identification.

Nycticeius

Vespertilio aenobarbus Temminck

1840.   Monographies de Mammalogie, 2:247, pl. 59, fig. 4.

RNH 17623: adult female; skin, skull removed; "Amérique méridionale"; collector and date of capture unknown. Holotype or syntype.

Skin.—Faded. Hair on dorsum and venter with basal half dark gray, distal half buff; tragus broad, bluntly rounded.
Skull.—Damaged. Third lower incisor slightly larger than first and second.

Remarks.—Following Miller and Allen (1928:200, 203), Cabrera (1958:102) and Hall and Kelson (1959:177) placed this name in synonymy with *Myotis albescens*, a species from which Husson (1962:217) reported that *Vespertilio aenobarbus* was specifically and generically distinct. Its generic affinity remains uncertain but it is probably with *Nycticeius*, and the specimen probably is not from South America.

Rhogeessa

Rhogeessa Alleni Thomas

1892.   Ann. Mag. Nat. Hist., ser. 6, 10:477. December.

BMNH (not numbered): adult female; in alcohol; Santa Rosalía, near Autlán, Jalisco, México; A. C. Buller; date of capture not specified. Holotype.

Skin.—Condition poor; faded; not measured.

Remarks.—The name Rhogeessa alleni is a patronym for Harrison Allen. Thomas based his description on two specimens of this species received from Dr. A. C. Buller. Because of its poor condition we did not attempt to measure the holotype; measurements recorded by Thomas are (holotype) head and body, 47; tail, 41; ear above head, 12.2; ear from notch, 16; tragus, inner margin, 7; forearm, 35; thumb, 5; met. III, 33.5; lower leg, 15.5; hind foot, 7.1; calcar, 15; (paratype) occiput-gnathion, 14.7; greatest breadth, 9.5; maxillary toothrow, 5.4.

Rhogeessa bombyx Thomas

1913.   Ann. Mag. Nat. Hist., ser. 8, 12:569. December.

BMNH 13.10.29.1: adult male; in alcohol, skull removed; Condoto, 300 ft., Chocó, Colombia; prepared by Dr. H. G. F. Spurrell; date of capture not specified. Holotype.

Remarks.—Thomas examined a single specimen.

Rhogeessa io Thomas

1903.   Ann. Mag. Nat. Hist., ser. 7, 11:382. April.

BMNH 94.9.25.1: adult male; skin and skull; Valencia [in state of Carabobo], Venezuela; A. Mocquerys, no. 196; November-December 1893. Holotype.

Skin.—Dorsum yellow-brown, hair with basal two-thirds yellow; venter yellowish.

Remarks.—Thomas gave forearm measurements for three paratypes in alcohol, two males and one female, and mentioned an additional skin from Bogotá. Thomas listed G. D. Child as the collector for these specimens.

Rhogeessa velilla Thomas

1903.   Ann. Mag. Nat. Hist., ser. 7, 11:383. April.

BMNH 99.8.1.5: adult male; skin and skull; Puná, Isla de Puná, Golfo de Guayaquil, Ecuador; Perry O. Simons, no. 43, measured 40, 35, 6, 7; 11 November 1898. Holotype.

Remarks.—Thomas examined a single specimen. "Shot flying in morning" is noted on the skin label. We were unable to distinguish between the skins of R. io and R. velilla; Thomas separated the two on the basis that the skull of the holotype of *R. velilla* was "quite without the marked 'helmet' found in all of the other forms" of *Rhogeessa.*

Lasiurus

Atalapha egregia Peters

1871.   Monatsb. Kön. preuss. Akad. Wiss. Berlin, 1870:912.

ZMB 3762: adult male; in alcohol; [state of] Santa Catarina, Brazil; H. Burmeister; date of capture not specified. Holotype.

Skin.—Apparently faded in part. Hair on top of head, shoulders, and upper portion of back with basal half reddish and distal half cream, on lower back and rump basal half blackish (or reddish black) and distal half reddish; hair on venter with basal one-half dark reddish brown or blackish, distal one-half reddish; uropatagium with reddish hair on proximal one-half of dorsal surface, hair on proximal edge only of ventral surface.

Remarks.—Peters apparently examined a single specimen.

Atalapha Frantzii Peters

1871.   Monatsb. Kön. preuss. Akad. Wiss. Berlin, 1870:908.

ZMB 2707: two adult females, in alcohol; Costa Rica; Hartzuil; date of capture not specified. Syntypes.

Skin.—Condition poor from loss of hair. Color darker than in ZMB 3451.

ZMB 3451: adult female; in alcohol; Brazil; Kickartz; date of capture not specified. Syntype.

Skin.—Hair on dorsum with basal band blackish, median band yellowish, distal band moderately dark reddish; color of ventral pelage similar to that of dorsum except distal red paler, perhaps once lightly frosted.

Remarks.—Peters based *Atalapha frantzii* on three specimens, two from Costa Rica obtained from Dr. von Frantzius and a third known only to be from Brazil. All three are identified as types. By implication, Miller (1897:111), apparently not realizing that one of Peters' syntypes was from Brazil, restricted the type locality to Costa Rica when he placed *Atalapha frantzii* in synonymy with *Lasiurus borealis mexicanus* (Saussure); subsequent authors have recognized Costa Rica as the type locality. At present, the name *frantzii* is applied subspecifically to populations of *Lasiurus borealis* from Chiapas south through Panamá (Jones *et al.*, 1977:24) into Colombia and Venezuela (Cabrera, 1958:113).

Atalapha pallescens Peters

1871.   Monatsb. Kön. preuss. Akad. Wiss. Berlin, 1870:910.

ZMB 595: adult female; in alcohol; Paramo de la Culata, Andes de Mérida, Venezuela; Karsten; date of capture not specified. Holotype.

Skin.—Faded. Other than fading, pelage color like that of North American specimens of *Lasiurus cinereus.*

Remarks.—Atalapha pallescens was based on a single specimen.

Dasypterus ega argentinus Thomas

1901.   Ann. Mag. Nat. Hist., ser. 7, 8:247. September.

BMNH 98.3.4.9: adult male; skin and skull; Goya, Corrientes, Argentina; R. Perrens, no. 58, measured 64, 43, 6, 14; 29 March 1896. Holotype.

Skin.—Dorsal pelage pale yellow, hair banded as in *Dasypterus ega panamensis* except color much paler, basal band of black narrower; ventral pelage as in *D. e. panamensis* except very pale.

Remarks.—"Killed in my sitting room" was noted on the skin label. Thomas examined another skin, in addition to the holotype, and mentioned a number of fluid-preserved specimens, which appeared "to be similarly coloured," from northern Argentina and the Bolivian and Paraguayan Chaco.

Dasypterus ega fuscatus Thomas

1901.   Ann. Mag. Nat. Hist., ser. 7, 8:246. September.

BMNH 99.9.6.5: adult of undetermined sex; skin and skull; Río Cauquita, 3000 ft. [3°27′N, 76°31′W (Aellen, 1970) in department of Valle], Colombia; J. H. Batty; 15 June 1898. Holotype.

Skin.—Dorsal pelage dull yellow with dull brown wash, hair banded (basal to distal) dark brown (about one-fourth length of hair), dull yellow, dull brown; ventral pelage dull, dirty yellow, hair with basal band pale brown, median band dirty yellow, becoming brownish toward tips; skin in general like that of *Dasypterus ega panamensis.*

Remarks.—Thomas described this subspecies on the basis of three specimens, all apparently collected by J. H. Batty (of Batty, Parish & Co.) at the same locality. The locality recorded on the specimen label is "Rio Cauqueta, Colombo, S.A., 3000 ft."; Thomas reported it as "Rio Cauquete, Cauca River, Colombia . . . . 1000 m."

Dasypterus ega panamensis Thomas

1901.   Ann. Mag. Nat. Hist., ser. 7, 8:246. September.

BMNH 0.7.11.1: adult male; skin and skull; Bogava [Bugaba], 250 m., Chiriquí, Panamá; H. J. Watson, no. 28, measured 77, 54, 9, 13; 8 October 1898. Holotype.

Skin.—Dorsum dull dark yellow, with dull brown wash, hair banded (basal to distal) black (approximately one-fourth length of hair), dull yellow, dull brown, dull yellow; venter dull yellow, basal half of hair black.

Remarks.—"Caught under leaves of bananas" was noted on skin label. Thomas examined a single specimen.

Dasypterus ega xanthinus Thomas

1897.   Ann. Mag. Nat. Hist., ser. 6, 20:544. December.

BMNH 98.3.1.14: adult male; skin and skull; Sierra Laguna, Baja California Sur, México; collection of W. W. Price, no. 397 (collected by D. Coolidge), measured 116, 48, 10, 16; 8 July 1896. Holotype.

Skin.—Pelage as in *Dasypterus ega panamensis* except richer, warmer yellow.

Remarks.—Thomas examined six specimens, all from Sierra Laguna at the southern tip of Baja California Sur.

Lasiurus caudatus Tomes

1857.   Proc. Zool. Soc., London, 15:42.

BMNH 44.10.9.7: adult of undetermined sex; skin, skull not removed; Pernambuco, Brazil; presented by J. P. G. Smith; date of capture not specified. Holotype.

Skin.—Color as in *Lasiurus ega xanthinus* except slightly duller; hair on dorsum with narrow basal band of dark gray.
Skull.—Condition undetermined.

Remarks.—Tomes mentioned two specimens in his description: the holotype, designated as "No. 1," and a second animal, also of unspecified sex, referred to only as "No. 2," and described as "a specimen in a bad state in spirit, from Chili." *Lasiurus grayi* Tomes (1857:40) was not found in the British Museum.

MOLOSSIDAE

Molossops

Dysopes abrasus Temminck

1827.   Monographies de Mammalogie, 1:232, pl. 21.

RNH 17374: adult female; skin and skull; Brazil; Maximilian, Prinz zu Wied-Neuwied; date of capture not specified. Holotype.

Skin.—Pelage reddish on dorsum, duller red on venter; overall appearance like that of TCWC 12433 from Perú.
Skull.—Rostrum and mandible only remain.

Remarks.—The name *Dysopes abrasus* long was misassigned to the genus *Eumops* (see Goodwin, 1960; Husson, 1962:243-246); *Dysopes abrasus* [= *Molossops (Cynomops) abrassus*] Temminck is a senior synonym for the name *Molossops brachymeles* (Peters). Although Temminck's single specimen was purported to be a juvenile, all other evidence in the Rijksmuseum indicates that RNH 17374 is the holotype of *Dysopes abrasus.*

Dysopes Temminckii Burmeister

1854.   Systematische ubersicht der Thiere Brasiliens . . . , 1:72.

ZMB 5458: adult female; in alcohol, skull not removed; Lagoa Santa, Minas Gerais, Brazil; H. Burmeister; date of capture not specified. Syntype.

Skin.—Condition poor, first preserved in alcohol, subsequently dried, and later returned to alcohol. Hair long, bicolored on back (basal half whitish, distal half pale brown).

Remarks.—Burmeister received two specimens of this species while in Lagoa Santa, but the whereabouts of a second specimen is unknown to us. Burmeister acknowledged Lund as the author of the name *Dysopes temminckii*, but we were unable to find Lund's published account.

Molossops mastivus Thomas

1911.   Ann. Mag. Nat. Hist., ser. 8, 7:113. January.

BMNH 10.11.10.3: adult male; skin and skull; Bartica Grove, lower [Río] Essequibo [in Guyana]; F. V. McConnell; date of capture not specified. Holotype.

Skin.—Dorsal pelage dark reddish brown, hair with basal portion slightly paler; venter paler than dorsum; ears and membranes blackish.
Skull.—Posterior part of skull much elevated; sagittal crest present.

Remarks.—A single specimen was examined, and, although McConnell's name appears on the label as collector, Thomas attributed the specimen's capture to Cozier and stated that it was presented to the British Museum by F. V. McConnell.

Molossops temminckii sylvia Thomas

1924.   Ann. Mag. Nat. Hist., ser. 9, 13:234. February.

BMNH 98.3.4.21: adult female; skin and skull; Goya, 600 ft., Corrientes, Argentina; R. Perrens, no. 140, measured 53, 24, 7, 12; 7 October 1896. Holotype.

Skin.—Occiput not much elevated; sagittal crest weak.

Remarks.—"Killed in hollow log" was noted on skin label. Thomas examined six specimens in addition to the holotype, but none was identified by number.

Molossus cerastes Thomas

1901.   Ann. Mag. Nat. Hist., ser. 7, 8:440. November.

BMNH 1.8.1.13: adult male; skin and skull; Villa Rica [probably Villarrica, province of Guairá], Paraguay; William Foster, no. 194, measured 90, 38, 9, ——, expanse 351; 22 January 1901. Holotype.

Skin.—Dorsal pelage dense, reddish gray in color, paler basally; venter same as dorsum in color; ears and membranes reddish brown to blackish.
Skull.—Sagittal crest present; back of skull moderately elevated.

Remarks.—"Caught in roof of house" was noted on the skin label. In addition to the holotype, Thomas examined a second male from Villa Rica, caught 26 January, and two females from Sapucay, caught on 9 and 10 June. None of these specimens was identified by number.

M [*olossus* (*Molossops*)]. *planirostris* Peters

1865.   Monatsb. Kön. preuss. Akad. Wiss. Berlin, p. 575.

ZMB 2513: adult male; in alcohol, skull removed; Cayenne [in French Guiana]; Schomburgk; date of capture not specified. Lectotype (here designated).

Skin.—Hair short, reddish brown.

Remarks.—Peters listed three specimens: one in the Berlin museum from British Guiana, one in the München museum from Barra do Rio Negro, and one in the Halle museum from Buenos Aires. Cabrera (1958:119), Hall and Kelson (1959:204), and Husson (1962:231-232) considered the locality first mentioned by Peters as the restricted type locality. ZMB 2513 is labeled (on the jar containing it) as having come from "Cayenne" (assumed by us to refer to the city, or its environs, by that name in French Guiana), but the locality entered in the Berlin museum catalogue was simply "Guiana." As far as we can determine, essentially all specimens taken in French Guiana through the first three-quarters of the nineteenth century were labeled as having come from Cayenne, and we assume that the cataloguer's entry of "Guiana" was an attempt, if any was intended, to clarify the label locality accompanying the specimen; Peter's reference to "British Guiana" is construed as a simple error based on the catalogue entry.

Because this specimen appears to be the only remaining syntype for *Molossops planirostris* and because it is the only one identifiable as being available to Peters at the time of its description, we herein designate number 2513, Zoologisches Museum der Humboldt-Universität zu Berlin, as lectotype for *Molossus (Molossops) planirostris* Peters.

Molossus planirostris paranus Thomas

1901.　Ann. Mag. Nat. Hist., ser. 7, 8:190. September.

BMNH 1.7.11.15: adult male; in alcohol, skull removed; Pará, Brazil; Goeldi Museum; date of capture not specified. Holotype.

Skull.—Posterior of skull not much elevated; sagittal crest weak. Skull similar in form and proportions to that of *Molossus temminckii sylvia.*

Remarks.—Thomas apparently examined a single specimen.

Tadarida

Dysopes auritus Wagner

1843.　Arch. Naturgesch., 9(1):368.

NMW (not numbered): adult female; skin, skull not removed; Cuyabá [Cuiabá, Mato Grosso], Brazil; Johann Natterer, no. 93; [15 March 1824]. Holotype.

Skin.—Faded somewhat. Dorsal pelage medium brown with slight reddish cast; venter pale brown with reddish cast; upper lips wrinkled; second phalanx of fourth digit short, as in *Tadarida macrotis.*

Skull.—Apparently entire.

Remarks.—Johann Natterer collected two specimens of *D. auritus*, both at Cuiabá, one in March, the other in September. One specimen was sent to Wagner, but Joseph Natterer (curator in charge of mammals at Vienna) asked that it be returned inasmuch as he had only one, or at most two specimens, of that and other species indicated on the invoice (apparently, Wagner returned this example and the others). No specimen was found at München, nor was a second one found at Vienna.

Dysopes gracilis Wagner

1843.   Arch. Naturgesch., 9(1):368.

ZSM 135: adult female; skin, skull not removed; Cuyabá [Cuiabá, Mato Grosso], Brazil; Johann Natterer; 12 July 1824. Syntype.

Skin.—Faded somewhat. Dorsal and ventral color moderately pale reddish brown.

Skull.—Apparently entire. First upper incisors parallel; lower incisors two in each half of jaw.

ZMB 2467: adult male; skin, skull not removed; Cuyabá [Cuiabá, Mato Grosso], Brazil; Johann Natterer; date of capture not specified. Syntype.

Skin.—Dorsal pelage dull reddish brown, hairs paler at base; venter paler than dorsum; ears only moderately enlarged; antitragus relatively long; upper lip wrinkled.

Skull.—Apparently entire. First upper incisors almost in contact, unusually long for a bat of *Tadarida macrotis* group.

Remarks.—An old label with ZSM 135 bears the number 102. According to the Berlin museum catalogue, ZMB 2467 is one of the original Natterer specimens on which the name *D. gracilis* was based. Wagner received at least two skins with skull in place from Joseph Natterer at Vienna, so we have assumed that the catalogue entry for ZMB 2467 is correct.

Dysopes multispinosus Burmeister

1861.   Reise durch die La Plata-Staaten . . . Argentinischen Republik. . . , 2:391.

ZMB 2614: two adult females; in alcohol, one with skull removed; Tucumán, Argentina; H. Burmeister; date of capture not specified. Syntypes.

Skin.—Both faded.

Remarks.—There are two specimens in the jar numbered 2614. According to the Berlin museum catalogue, these specimens are from the "Hall[e] Mus[eum]" in Germany.

D[*ysopes*]. *Naso* Wagner

1840.   Die Säugthiere . . . von Schreber, Supplementband, 1:475.

ZMB 2464: adult male; skin, skull not removed; Ypanema [Ipanema, in state of São Paulo], Brazil; Johann Natterer; date of capture not specified. Syntype.

Skin.—Faded somewhat. Dorsal pelage dull brown with slight grayish cast; venter paler than dorsum.

ZSM 130: adult female; skin, skull not removed; Ytararé [Itararé, São Paulo], Brazil; Johann Natterer; 30 August 1820. Syntype.

Skin.—Faded somewhat. Dorsal pelage medium reddish brown; venter paler than dorsum. *Skull.*—Incomplete.

ZSM 134: adult female; skin, skull not removed; Ypanema [Ipanema, São Paulo], Brazil; Johann Natterer; 1819. Syntype.

Skin.—Faded somewhat. Dorsal pelage medium reddish brown; venter paler than dorsum. *Skull.*—Apparently entire. First upper incisors converging.

Remarks.—The original entry in the Berlin museum catalogue indicates that ZMB 2464 was received as a type specimen of *Dysopes naso.* The original skin labels for ZSM 130 and 134 were written by Johann Natterer, "No. 60 / Ytarare / 30 Aug / Foemina" and "N. 60 / Ypanema / 1819 / Foemina," respectively.

Dysopes nasutus Temminck

1827.   Monographies de Mammalogie, 1:233, pl. 24, figs. 2, 3.

RNH 17575: adult of undetermined sex; skin, skull not removed; Brazil; Maximilian, Prinz zu Wied-Neuwied; date of capture not specified. Syntype.

RNH 17576: adult female; skeleton; Brazil; Johann Natterer; date of capture not specified. Syntype.

Remarks.—Of the several specimens apparently available to Temminck, two were found. The skeleton is the one figured in pl. 24, fig. 2. Temminck's description was intended as a reassignment of *Molossus nasutus* Spix (a species of the genus now recognized as *Promops*) to the genus *Dysopes.* Temminck also placed the name *D. nasutus* in synonymy with *Nyctinomus brasiliensis,* where it belongs.

M[olossus]. mexicanus Saussure

1860.   Rev. Mag. Zool. Paris, ser. 2, 12:283, pl. 15, fig. 2, 2a.

ZMB 2589: adult male; in alcohol; [Cofre de Perote, 13,000 ft., Veracruz], México; Saussure; date of capture not specified. Lectotype (here designated).

Remarks.—This specimen, labeled "cotype" (=syntype), was obtained through an exchange with the Paris museum. Inasmuch as it was collected by Saussure, we believe that it must be one of the original specimens on which he based the name *Molossus mexicana.* The number of specimens was not stated, but Saussure referred to one from Cofre de Perote, 13,000 ft., and others from Ameca, at the foot of Popocatepetl, 8,500 ft. Shamel (1931:5), thinking that Saussure's Amecan specimens had come from the Jaliscan town by that name, restricted the type locality to Ameca, Jalisco, because the United States National Museum had three specimens from the immediate vicinity of that locality. Notwithstanding, the Ameca to which Saussure referred was "au pied du Popocatepetl," probably in the vicinity of the town now known as Amecameca de Juarez, in the state of México. Hall and Kelson (1959:206) listed Cofre de Perote as the type locality, and we restrict it herewith to that place, a volcano, the crest of which is approximately 23 km. WSW Jalapa Enríquez, Veracruz. The Berlin museum specimen appears to be the only remaining syntype of *Molossus mexicanus*, and, herewith, we designate it to be the lectotype for this name.

Molossus rugosus d'Orbigny

1837.   Voyage dans l'Amérique Méridionale . . . années 1826 . . . 1833, 9(Atlas Zoologique), pl. 10, figs. 3-5.

MNHN 795J: adult of undetermined sex; skin, skull not removed; [probably Tucumán], Corrientes, Argentina; d'Orbigny; date of capture not specified. Holotype or syntype.

Remarks.—This specimen was not identified as a type by Rode (1941), but available evidence indicates that it is one of d'Orbigny's specimens and therefore a type for *Molossus rugosus*. There is a second specimen (MNHN 795K, adult female, skin, skull not removed, taken by Gay in Chile) labeled *Molossus rugosus*, but we do not know whether it was available to d'Orbigny. According to Sherborn and Griffin (1934), the date of publication for pl. 10 of the Atlas Zoologique was 1837.

Nyctinomus brasiliensis I. Geoffroy St.-Hilaire

1824.   Ann. Sci. Nat. Paris, 3:337, pl. 22.

MNHN 800: adult of undetermined sex; skin, skull not removed; Brazil [Curityba, Paraná, by restriction (Shamel, 1931:4)]; Auguste de St.-Hilaire; date of capture not specified. Syntype (no. 230).

MNHN 801: juvenile of undetermined sex; skin and skull (no. A.6826, comparative anatomy laboratory); Brazil; Auguste de St.-Hilaire; date of capture not specified. Syntype (no. 230*a*).

MNHN 802: adult of undetermined sex; skin, skull not removed; Brazil; Auguste de St.-Hilaire; date of capture not specified. Syntype (no. 230*b*).

MNHN 803: adult of undetermined sex; skin, skull not removed; Brazil; Auguste de St.-Hilaire; date of capture not specified. Syntype (no. 230*c*).

Remarks.—The type numbers are those assigned by Rode (1941).

Nyctinomus macrotis Gray

1839.   Ann. Nat. Hist., 4:5. September.

BMNH (not numbered): adult female; in alcohol, skull removed; Cuba; W. S. MacLeay; date of capture not specified. Holotype.

Skin.—Condition fair, some fading and some loss of hair.
Skull.—Slightly damaged.

Remarks.—MacLeay reported that this specimen came from the hollow of a tree in the interior of the island. The title page of vol. 4 of the Annals of Natural

History is dated 1840 but the number of which Gray's paper is a part was published in September 1839.

Nyctinomus megalotis Dobson

1876.   Proc. Zool. Soc., London, p. 728.

BMNH (not numbered): adult male; in alcohol, skull removed; Surinam; collector and date of capture unknown. Holotype.

Skin.—Condition fair; some fading, some loss of hair.
Skull.—Removed, but uncleaned, in fluid with body, and damaged; mandible intact.

Remarks.—Dobson examined a single specimen. Although no number is affixed to the holotype, the British Museum catalogue entry 64.6.18.6 (a Surinam specimen obtained from Peters at the Berlin museum) possibly refers to the holotype for *N. megalotis*. Dobson (1876, 1878) made no mention of how the holotype came into the possession of the British Museum.

Nyctinomus musculus Peters

1861.   Monatsb. Kön. preuss. Akad. Wiss. Berlin, p. 149.

ZMB 2482: adult male; skin, skull not removed; Cuba; Gundlach; date of capture not specified. Syntype.

Remarks.—There are four additional syntypes, two adult males and two adult females, all numbered ZMB 2457 and in alcohol with skulls not removed. All five syntypes were taken in Cuba by Gundlach, to whom Peters credited the name.

Mormopterus

Nyctinomus kalinowskii Thomas

1893.   Proc. Zool. Soc., London, p. 334, pl. 29, fig. 10.

BMNH 94.8.6.7: adult female; in alcohol, skull removed; central part of Perú; J. Kalinowski; date of capture not specified. Holotype.

Skull.—Much flattened dorsoventrally; dental formula 1/2, 1/1, 1/2, 3/3; first upper incisor anteroposteriorly flattened; upper canine not in contact with upper premolar; anterior emargination in palate broad and first upper incisors convergent.

Remarks.—Thomas identified the collector as M. Kalinowski; the only label now affixed to the holotype was written at the British Museum after the specimen was acquired from the Warsaw museum.

Eumops

Dysopes glaucinus Wagner

1843.   Arch. Naturgesch., 9(1):368.

NMW (not numbered): juvenile of undetermined sex, probably male; skin, skull not removed; Cuyabá [Cuiabá, Mato Grosso], Brazil; Johann Natterer; date of capture not specified. Holotype.

Skin.—Faded somewhat. Dorsum pale brown, hairs white at base; venter pale gray brown, hairs white at base, tipped with white; ears relatively short; chiropatagium with relatively dense strip of hair on proximal portion of surface from elbow caudad.

Skull.—Apparently entire.

Remarks.—A label pinned to the specimen's back, and in Fitzinger's handwriting, bears the notation "*D. glaucinus.* Natt. / Cuyaba. Mas. 8." The original label is missing. Hall and Kelson (1959:213) stated that the type was "purportedly from Cuyaba"; all indications are that it was. Wagner examined a single specimen.

Dysopes longimanus Wagner

1843.   Arch. Naturgesch, 9(1):367.

ZSM 56: adult female; skin, skull not removed; Villa Maria, Caiçara, Barra do Rio Negro [vicinity of Manaus, Amazonas], Brazil; Johann Natterer; February 1826. Syntype.

ZSM 60: adult male; skin, skull not removed; Villa Maria, Caiçara Barra do Rio Negro [vicinity of Manaus, Amazonas], Brazil; Johann Natterer; January 1826. Syntype.

Skin (ZSM 56 and 60).—Faded somewhat. Dorsal pelage reddish brown almost to base, hairs whitish at base; ventral pelage slightly paler than dorsal; ears relatively small.

Skull (ZSM 56 and 60).—Apparently entire.

Remarks.—Two labels are affixed to each of the two syntypes: those for ZSM 56 are inscribed "No. 23b / Caicara / Febr. 826 / Foemina" (original label, written by Johann Natterer) and "*Dysopes longimanus* Wagn. / E Mus Vindob. / Bras." (old green label); for ZSM 60, "No. 23b / Caicara / Janer. 826 / Mas" (original label, written by Johann Natterer, original number emended to 23b) and "*Dysopes longimanus* Wagn. / 1844 / Bras." (old green label).

Dysopes (Molossus) gigas Peters

1864.   Monatsb. Kön. preuss. Akad. Wiss. Berlin, p. 381.

ZMB 2474: juvenile male; skin, skull not removed; Rio Negro [in state of Amazonas], Brazil; Johann Natterer; date of capture not specified. Holotype or syntype.

Skin.—Faded. Hair on dorsum with basal two-thirds cream, distal one-third pale reddish gray; venter paler than dorsum; ears large, broad; antitragus broadly rounded.

Remarks.—This name, apparently a synonym of *Eumops perotis* (Schinz, 1821), seems to have been overlooked by subsequent authors.

Eumops dabbenei Thomas

1914.   Ann. Mag. Nat. Hist., ser. 8, 13:480. May.

BMNH 14.4.4.8: adult female; in alcohol, skull removed; [province of] Chaco, Argentina; collector and date of capture unknown. Holotype.

Remarks.—This specimen, "caught on board S.S. *Obidense*" (as noted on the specimen label) was obtained from the Buenos Aires museum. *Eumops dabbenei* is a patronym for Dr. R. Dabbene, a former Conservator of Zoology at the Buenos Aires National Museum.

Eumops delticus Thomas

1923.   Ann. Mag. Nat. Hist., ser. 9, 12:341. September.

BMNH 23.8.9.7: adult female; skin and skull; Caldeirão, Ilha de Marajó [at mouth of Rio Amazonas, Pará], Brazil; W. Ehrhardt, no. 88, measured 68, 41, 10, 19; 14 January 1923. Holotype.

Skin.—Dorsal pelage dark gray brown, hair pale gray at base; ventral pelage gray brown with gray wash, hair with basal band almost white, median band gray brown, distal one pale gray.
Skull.—Basisphenoid pits well defined, perhaps exceptionally so; occiput visible from above.

Remarks.—Thomas examined a single specimen.

Eumops geijskesi Husson

1962.   The bats of Suriname, p. 246, figs. 1a, 36c, pl. 28.

RNH 12943: adult female; in alcohol, skull removed; Surinam; W. J. Bresser; 1862. Holotype.

Skin.—Hair on back approximately 8 millimeters in length, dark reddish brown from base to tip; venter slightly paler than dorsum; length of ear from meatus approximately 15.
Skull.—First upper premolar minute, crowded toward lateral side of toothrow; basisphenoid pits separated by broad ridge; sagittal crest low; lambdoidal crest moderately developed, occiput visible from above.

Remarks.—Jentink (1888:200) assigned the holotype and three paratypes to *Molossus rufus* Geoffroy; Jentink's specimens *h* and *i* are skins; *g* (holotype) and *j* are in alcohol. Paratypes RNH 12944 and 12945 (subadults, male and female, respectively) are skins and skulls; both have white hair on underside of chiropatagium next to body. Paratype 12946 is a juvenile with milk teeth; permanent teeth were erupting at time of capture.

Eumops patagonicus Thomas

1924.   Ann. Mag. Nat. Hist., ser. 9, 13:234. February.

BMNH 23.12.12.18: adult female; in alcohol, skull removed; [province of] Chubut, Argentina; collector and date of capture unknown. Holotype.

Remarks.—This specimen was obtained from the Buenos Aires museum; its museum number there was 4068. Thomas apparently examined a single specimen.

Molossus ferox Peters

1861.   Monatsb. Kön. preuss. Akad. Wiss. Berlin, p. 149.

ZMB 2587: adult female; skin, skull not removed; Cuba; Gundlach; date of capture not specified. Syntype.

ZMB 2865: adult, probably female; skin, skull not removed; Cuba; Gundlach; date of capture not specified. Syntype.

ZMB 2871: adult male; skin, skull not removed; Cuba; Gundlach; date of capture not specified. Syntype.

ZMB 2981: adult female; skin and skull; Cuba; Gundlach; date of capture not specified. Syntype.

Skin.—Pelage of all four faded somewhat. Hair on back whitish at base, becoming buff then reddish brown distally (dorsum of ZMB 2871 has a gray cast, 2865 is brown with a dull reddish cast); ventral color paler, tends to have a pale grayish cast, with hairs white basally; ears only moderately enlarged for *Eumops*; antitragus typical for *Eumops*.

Molossus maurus Thomas

1901.   Ann. Mag. Nat. Hist., ser. 7, 8:141. August.

BMNH 1.6.4.34: adult male; skin and skull; Kanuku Mts., 240 ft., British Guiana [Guyana]; J. J. Quelch (prepared and presented by F. V. McConnell), measured 63, 50.5, 11, 19, expanse 388; 11 December 1900. Holotype.

Skin.—Pelage dark reddish brown above and below, hair on back bicolored (narrow whitish basal band), hair on venter paler at base.
Skull.—Single upper premolar, in contact with upper canine; occiput visible from above.

Remarks.—Thomas apparently examined a single specimen.

Promops nanus Miller

1900.   Ann. Mag. Nat. Hist., ser. 7, 6:470. November.

BMNH 0.7.11.99: adult male; skin and skull; Bogava [Bugaba], 250 m., Chiriquí, Panamá; H. J. Watson, no. 56, measured 61, 34, 6, 13; 7 October 1898. Holotype.

Remarks.—The holotype was "caught under roof of house," as noted on the skin label. Bugaba is spelled Bogava on the label and in Miller's description. Miller stated that he examined a second skin (female) from Chiriquí but did not identify it by number.

Promops Trumbulli Thomas

1901.   Ann. Mag. Nat. Hist., ser. 7, 7:190. February.

BMNH 99.11.2.1: subadult, probably female; skin and skull; Pará, Brazil; J. Trumbull; 7 June 1898. Holotype.

Skin.—Dorsal pelage medium gray with a slight reddish cast, hair becoming pale gray to almost white at base; venter pale gray with slight buff cast, hairs becoming paler to almost white at base; ears large and broad, with area of thick stiff hairs (much thicker than is common for specimens of *Eumops auripendulus*) approximately 2 millimeters long on underside.
Skull.—Rostrum elongate, narrow (anterior portion not much wider than postorbital constriction); basisphenoid pits well developed, divided by narrow partition.

Remarks.—The catalogue entry for 99.11.2.1 indicates that it was a skin only, but there is a skull with the same number, also marked "type." Thomas compared his single specimen of *Promops trumbulli* with *Eumops perotis,* to which he stated that the former was "closely allied . . . but with smaller ears, smaller tragus, and much smaller and lighter teeth."

Promops

Molossus Fosteri Thomas

1901.   Ann. Mag. Nat. Hist., ser. 7, 8:438. November.

BMNH 1.8.1.17: adult male; skin and skull; Villa Rica [probably Villarrica, province of Guairá], Paraguay; William Foster, no. 1234; 26 February 1901. Holotype.

Skin.—General appearance similar to that of TCWC 12838; we are unable to distinguish between the two.

Remarks.—Thomas examined five additional specimens, one male and four females, none of which was identified by number. These were captured at two localities, Villa Rica, Guairá, and Sapucay, Paraguarí.

Molossus nasutus Spix

1823.   Simiarum et vespertilionum brasiliensium species novae . . . , p. 60, pl. 35, fig. 7.

ZSM 136: adult of undetermined sex; skin and skull; Rio São Francisco [in state of Bahia], Brazil; apparently Spix; date of capture not specified. Holotype or syntype.

Skin.—Faded somewhat. Dorsal pelage reddish brown, hairs whitish at base; ventral pelage much paler than dorsal.
Skull.—As in TCWC 12838, except smaller, sagittal and lambdoidal crests slightly weaker.

Remarks.—An old green label bears the notation: "*Molossus* [*Dysopes*] *nasutus* Spix / original / Brazil." With regard to the type locality, Spix stated "reperitur sub tectis domuum prope flumen St. Francisci."

Promops ancilla Thomas

1915.   Ann. Mag. Nat. Hist., ser. 8, 16:63. July.

BMNH 6.5.8.4: adult male; in alcohol, skull removed; Cachi, 2500 m., Salta, Argentina; J. Steinbach; 15 April 1905. Holotype.

Skin.—Similar to that of TCWC 12838 in color but paler and with slight reddish cast, hair whitish at base.
Skull.—Like that of TCWC 12838, except smaller.

Remarks.—In addition to the holotype, Thomas examined a specimen (not identified by number) from Tucumán, Corrientes, Argentina.

Promops centralis Thomas

1915.   Ann. Mag. Nat. Hist., ser. 8, 16:62. July.

BMNH 94.2.5.4: adult female; skin and skull; "N. Yucatán," México; [collected by G. F. Gaumer, presented by] Osbert Salvin; date of capture not specified. Holotype.

Skin.—Dorsal pelage dark reddish brown, hairs pale (almost gray) at base.
Skull.—Like that of TCWC 12838, except larger.

Remarks.—Thomas examined two additional specimens but did not identify them by number. Although the pelage is somewhat darker, the holotype for *Promops centralis* is similar to that of *P. occultus.*

Promops davisoni Thomas

1921.   Ann. Mag. Nat. Hist., ser. 9, 8:139. July.

BMNH 21.5.21.1: adult male; skin and skull; Chosica, 2700 ft., Lima, Perú; J. F. Davison, no. 207, measured 72, 55, 10, 13; 3 March 1921. Holotype.

Skin.—Like that of TCWC 12838.
Skull.—Like that of TCWC 12838.

Remarks.—Thomas examined one additional specimen but did not identify it by number.

Promops occultus Thomas

1915.   Ann. Mag. Nat. Hist., ser. 8, 16:62. July.

BMNH 2.11.7.24: adult female; skin and skull; Sapucay [in department of Paraguarí], Paraguay; William Foster, no. 714, measured 86, 56, 11, 17, expanse 393; 6 March 1902. Holotype.

Skin.—Dorsal pelage dark reddish brown, hairs paler (becoming gray) at base; ventral pelage dull reddish gray, hairs paler at base.
Skull.—Like that of TCWC 12838, except larger.

Remarks.—"Caught in palm tree" was noted on the skin label. Thomas also examined two fluid-preserved specimens and another skin and skull but did not identify these by number. There is a discrepancy in the BMNH catalogue number (2.4.11.24) cited by Thomas and that appearing on the specimen now labeled "Type." BMNH 2.4.11.24 (the 24th specimen cataloged on 11 April 1902) is an obvious error; the collector's field number, 714, was reported correctly. Except for its larger size and darker pelage, the holotype of *Promops occultus* is similar to TCWC 12838.

Molossus

Dysopes albus Wagner

1843.   Arch. Naturgesch., 9(1):368.

NMW (not numbered): adult male; skin, skull not removed; "En. d. Gama," Mato Grosso; Johann Natterer; 26 July 1826. Holotype.

Skin.—Pelage white; membranes blackish.

Remarks.—This specimen no longer bears a label but clearly it is the one that Wagner referred to as "D[ysopes]. supra subtusque albidas; patagiis nigricantibus." *D. albus* is a junior synonym of *Molossus rufus* É. Geoffroy St.-Hilaire, 1805.

Dysopes alecto Temminck

1827.   Monographies de Mammalogie, 1:231, pls. 20, 23, figs. 23-26.

RNH 13023: adult of undetermined sex; skin, skull removed; interior of Brazil; Maximilian, Prinz zu Wied-Neuwied; date of capture not specified. Holotype.

Skin.—Pelage dark reddish brown above, slightly paler below; hair on back short, uni-colored.

Skull.—Occiput damaged. Skull like that of *Molossus rufus* É. Geoffroy St.-Hilaire.

Remarks.—The skull is the one figured on pl. 23.

Molossus Burnesi Thomas [printer's error for *Molossus Barnesi*]

1905.   Ann. Mag. Nat. Hist., ser. 7, 15:584. June.

BMNH 5.1.8.7: adult female; in alcohol, skull removed; Cayenne [in French Guiana]; W. Barnes; date of capture not specified. Holotype.

Skull.—Like that of TCWC 12585 but a little smaller.

Remarks.—As pointed out by Cabrera (1958:129), the spelling of *M. Burnesi* was an obvious typographical error; Thomas intended the name as a patronym for W. Barnes and spelled it "*M. Barnesi*" on p. 585. However, Miller (1913:91), as the first reviser, spelled the name *Molossus burnesi*, which Husson (1962:259) argued, according to Article 32(b) of the 1961 International Code of Zoological Nomenclature (p. 35), should be accepted as the correct original spelling. Notwithstanding, *M. burnesi*, because it is an inadvertent error [Art. 32a (ii)], has no status as an incorrect original spelling and is here corrected under Art. 32c to *Molossus barnesi*.

Thomas apparently examined a single specimen.

Molossus fluminensis Lataste

1891.   Ann. Mus. Civ. Stor. Nat. Giacomo Doria, ser. 2, 10:658. 11 April.

BMNH (not numbered): adult male; in alcohol, skull removed; Rio [de Janeiro, Brazil]; Naegeli; 1873. Holotype or syntype.

Skull.—Like that of *Molossus rufus* É. Geoffroy St.-Hilaire.

Remarks.—Lataste was Director of the Museo Nacional de Chile.

Molossus fuliginosus Gray [replaced by the substitute name *Molossus milleri* Johnson, 1952]

1838.   Mag. Zool. Bot., London, 2:501. February.

BMNH 28a: adult of undetermined sex; skin, skull not removed; "Bermuda"; T. Cottle; date of capture not specified. Paralectotype.

Skin.—Hair white basally, dull reddish brown distally; dorsum darker than venter.

Remarks.—Johnson (1952) proposed the substitute name *Molossus milleri* for *Molossus fuliginosus* Gray, 1838, which is preoccupied by *Molossus fuliginosus* Cooper, 1837. The specimen is now labeled as *Molossus milleri* Gray in the British Museum collection. Gray (1838) indicated that the origin of the specimens in his possession was unknown, but later Gray (1843:35) listed, in the following order, three specimens of *M. fuliginosus,* one from Bermuda (presented to the Museum by Thomas Cottle), one from Jamaica (presented by Thomas Bell), and one from Portobello (presented by the Lords of the Admiralty); the first, a skin, the other two, in spirits. Dobson (1878:413) included these three in his list of *Molossus obscurus* but identified as type (= lectotype by our interpretation) the specimen from Jamaica; Miller (1913:90) followed Dobson, as did Hall and Kelson (1959:216). The fluid-preserved specimens, both females, from Jamaica and Portobello were not found, although they may yet exist in the British Museum. Gray's (1839:7) second reference to a small throat pouch ("gland on throat very small, rudimentary") suggests that he examined one or more female specimens preserved in fluid. The specimen purported to be from Bermuda is marked "Type."

Molossus longicaudatus É. Geoffroy St.-Hilaire

1805.   Ann. Mus. Nat. Hist. Nat. Paris, 6:155.

MNHN 792: adult, probably female; skin and skull; locality, collector, and date of capture unknown. Holotype (also, paralectotype of *Vespertilio molossus* Pallas).

Skin.—Faded. Individual hairs bicolored, basal band whitish; general appearance similar to that of TCWC 12585.

Skull.—Smaller than those of MNHN A.419/225♂ (syntype for *Molossus obscurus* É. Geoffroy St.-Hilaire, 1805) and TCWC 12585 ♂ (differences probably due to sex).

Remarks.—This specimen was listed by Rode (1941) as type number 226. The locality from which it came or how it was obtained could not be determined from Paris museum records. É. Geoffroy St.-Hilaire's (p. 155) statement "elle est décrite par M. Daubenton sus le nom de *mulot volant*" is incorrect; Daubenton (in De Buffón and Daubenton, 1763:87) referred only to "No. DCDVI . . . (*pl. xix, fig. 1)*" by the name mulot-volant. É. Geoffroy St.-Hilaire's *M. longicaudatus* is Daubenton's "No DCDVII . . . (*pl. xix, fig. 2).*" The origin of Daubenton's mulot-volant was Martinique, but that of the specimen illustrated in pl. 19, fig. 2, apparently was unknown to Daubenton. We believe, as did Husson (1962), that *M. longicaudatus* is a junior objective synonym of *Vespertilio molossus minor* Kerr, 1792 (also based on De Buffon and Daubenton, 1763, pl. 19, fig. 2) and came from some place other than Martinique. See Husson (1962:256-258) for an extensive discussion of the names resulting from *Vespertilio molossus* Pallas, 1766, a composite that included the two bats figured by Daubenton and *Nyctinomus macrotis* Gray, 1839.

The lectotype of *Vespertilio molossus* (see Husson, 1962:257), which also is the holotype of *Vespertilio molossus major* Kerr, 1792, and *Molossus fusciventer* É. Geoffroy St.-Hilaire, 1805, could not be found at the Paris museum. É. Geoffroy St.-Hilaire's statement, "telles sont les quatre espèces [*Molossus rufus, M. ater, M. obscurus,* and *M. longicaudatus*] existantes au Muséum d'histoire naturelle; le second mulot volant [actually the only one so identified by Daubenton] que M. Daubenton a décrit dans *l'Histoire générale et particulière,* est l'espèce suivant," preceding his description of *Molossus fusciventer* indicates that the specimen from Martinique was not at the Paris museum in 1805, although it had been in the Cabinet du Roi when Daubenton described it as mulot-volant. The Cabinet du Roi became the Muséum National d'Histoire Naturelle, Paris, following the French Revolution.

Molossus milleri Johnson [substitute name for *Molossus fuliginosus* Gray]

1952.   Proc. Biol. Soc. Washington, 65:197. 5 November 1952.

BMNH 28a: adult of undetermined sex; skin, skull not removed; "Bermuda"; Thomas Cottle; date of capture not specified. Paralectotype.

Remarks.—See the account for *Molossus fuliginosus* for additional information.

Molossus obscurus É. Geoffroy St.-Hilaire

1805.   Ann. Mus. Nat. Hist. Nat. Paris, 6:155.

MNHN A.419/225-225*a*: two adult males; in alcohol, one with skull removed; Antilles [Martinique, by restriction (Husson, 1962:258)]; Moreau de Jonnes; date of capture not specified. Syntypes.

Skin.—Hair lost on lower back. Pelage red, hairs with basal band white.
Skull.—Similar to that of TCWC 12585 ♂; smaller than those of TCWC 12481 and 12576; the skulls of A.419 and TCWC 12585 resemble each other more closely than do their skins.

Remarks.—There are two specimens in the bottle labeled by Rode: "Type / no. 225-225a"; the skull is removed from 225. Another label, much older than the one affixed to the jar by Rode, bears the notation: "*Molossus obscurus* Geoff. / Type A.419 / Morreau Jonnes Antilles." The specimen label once affixed to 225 is now missing. Although É. Geoffroy St.-Hilaire (p. 155) gave no locality for the specimens on which he based the name *M. obscurus,* there is no reason to believe that they did not come from the Antilles as stated on the label. Subsequent authors were led to believe that the type locality was Surinam because Temminck's (1826:236-237) redescription of *M. obscurus* was based on material from Surinam.

Molossus obscurus currentium Thomas

1901.   Ann. Mag. Nat. Hist., ser. 7, 8:437. November.

BMNH 98.3.4.28: adult male; skin and skull; Goya [in province of Corrientes], Argentina; R. Perrens; 1 October 1896. Holotype.

Skin.—Hair bicolored on dorsum and venter, with white basal and gray distal bands.

Skull.—Slightly more elongate, with occiput more elevated, than normal for species of *Molossus*; otherwise, skull like that of TCWC 12481.

Molossus rufus É. Geoffroy St.-Hilaire

1805.   Ann. Mus. Nat. Hist. Nat. Paris, 6:155.

MNHN A.428/224-224a: two adult males; in alcohol, one (224) with skull removed; Amérique [Cayenne, French Guiana, by restriction (Miller, 1913)]; collector and date of capture unknown. Lectotype (A.428/224) here designated.

Skin.—Condition of both poor, much of hair lost. Pelage dark red, hair without bands.

Skull (224).—Basisphenoidal pits shallow, small, widely separated by low broad ridge.

Remarks.—Both specimens are in the same jar, to which two labels are glued: 1) "Type / no. 224-224a"; 2) "*Molossus rufus* Geoff. / Type A.428 / Amerique." The first label probably was put on by Rode, but the second label is much older.

Because of Goodwin's (1960) misinterpretation of É. Geoffroy St.-Hilaire's description of *Molossus rufus*, we designate specimen A.428/224, Muséum National d'Histoire Naturelle, Paris, as lectotype for the name *Molossus rufus* É. Geoffroy St.-Hilaire.

Molossus tropidorhynchus Gray

1839.   Ann. Nat. Hist., 4:6. September.

BMNH 54.12.27.6: adult male; in alcohol, skull removed; Cuba [probably Havana]; W. S. MacLeay; date of capture not specified. Holotype.

Skin.—Found but not examined.

Skull.—Slightly damaged. Like that of TCWC 12585.

Remarks.—MacLeay's statement (in Gray, 1839:6), "very common in the city of Havana," suggests to us that this specimen came from Havana. Gray examined a single specimen.

V[espertilio]. mol[ossus]. minor Kerr

1792.   The animal kingdom . . . , p. 97.

MNHN 792: adult, probably female; skin and skull; locality, collector, and date of capture unknown. Holotype.

Remarks.—This name was based on the lesser of two bats figured in De Buffon and Daubenton (1763, pl. 19). For additional details, see account for *Molossus longicaudatus* É. Geoffroy St.-Hilaire, 1805, a junior objective synonym of *Vespertilio molossus minor* Kerr. Miller's (1913:90) statement that Kerr's *V. m. minor* was "probably from Martinique" appears to have been predicated on Daubenton's mulot-volant (= *Vespertilio molossus major* Kerr, the larger of the two bats figured in De Buffon and Daubenton, 1763, pl. 19) having come from Martinique.

Acknowledgments

For the opportunity to examine specimens in their charge and for their many courtesies, hospitality, and assistance in locating specimens and museum records, we wish to express our sincere gratitude to Gordon B. Corbet, John Edwards Hill, and John Hayman, British Museum (Natural History); Francisco Bernij and Luis Blas, Museo Nacional de Ciencias Naturales; François de Beaufort and Jean Dorst, Muséum National d'Histoire Naturelle; Kurt Baur, Naturhistorisches Museum Wien; Alf G. Johnels, Naturhistoriska Riksmuseum; A. M. Husson, Rijksmuseum van Natuurlijke Historie; Heinz Felten, Natur-Museum und Forschungs-Institut Senckenberg; A. Kleinschmidt, Staatliches Museum für Naturkunde in Stuttgart; Georg H. W. Stein, Institut für Spezielle Zoologie und Zoologisches Museum der Humboldt-Universität zu Berlin; H. Schliemann, Zoologisches Staatsinstitut und Zoologisches Museum, Hamburg; and Theodor Haltenorth and Ingird Weigel, Zoologisches Staats-Samlung München. We extend our special thanks to John Edwards Hill and to M. J. Rowlands and his library staff at the British Museum (Natural History) for their valuable bibliographic assistance. William B. Davis, Alfred L. Gardner, Karl F. Koopman, and Don E. Wilson critically read the manuscript, and we appreciate their helpful comments.

This investigation was initiated as a segment of Project 1556 (Chiroptera of the Central American Core; their systematics, natural history, and ecological distribution) of the Texas Agricultural Experiment Station and was supported in part by National Science Foundation grant GB 3201.

Literature Cited

Aellen, V.   1970.   Catalogue raisonné des chiroptères de la Colombie.   Rev. Suisse Zool., 77:1-37.

Anthony, H. E.   1918.   The indigenous land mannals of Porto Rico, living and extinct.   Mem. Amer. Mus. Nat. Hist., 2:331-435.

Avila-Pires, F. D. de.   1965.   The type specimens of Brazilian mammals collected by Prince Maximilian zu Wied.   Amer. Mus. Novit., 2209:1-21.

Buffon, [G. L. le Clerc comte] de, and [L. J. M.] Daubenton.   1763.   Histoire naturelle, générale et particuliére, avec la description du Cabinet du Roi.   Paris, 10:xii + 368 pp.

Burmeister, H.   1854.   Systematische ubersicht der Thiere brasiliens welche mährend einer Reise durch die Provinzen von Rio de Janeiro und Minas geraës. . . .   Säugethiere (Mammalia).   Georg Reimer, Berlin, 1:x + 1-341.

Cabrera, A.   1958.   Catálogo de los mamíferos de America del Sur.   Rev. Mus. Argentino Cien. Nat. "Bernardino Rivadavia," Cien. Zool., 4:i-xxii + 1-307.

Carter, D. C.   1966.   A new species of Rhinophylla (Mammalia, Chiroptera: Phyllostomatidae) from South America.   Proc. Biol. Soc. Washington, 79:235-238.

Davis, W. B.   1973.   Geographic variation in the fishing bat, Noctilio leporinus.   J. Mamm., 54:862-874.

Dobson, G. E.   1876.   A monograph of the group Molossi.   Proc. Zool. Soc. London, 1876:701-735.

——. 1878. Catalogue of the chiroptera in the collection of the British Museum. British Museum, London, xlii + 1-567 + 30 pls.

GENOWAYS, H. H., AND R. J. BAKER. 1972. Stenoderma rufum. Mammalian Species, 18:1-4.

GEOFFROY ST.-HILAIRE, É. 1810. Sur les Phyllostomes et les Mégadermes, deux genres de la famille des chauve-souris. Ann. Mus. Nat. Hist. Nat. Paris, 15:157-198 + pls. ix-xii.

——. 1818a. Sur de nouvelles chauve-souris, sous le nom de glossophages. Mém. Mus. Hist. Nat. Paris, 4:411-418 + pls. xvii-xviii.

——. 1818b. Description des mammifères qui se trouvent en Égypte. Pp. 99-166, in Description de l'Égypte, ou recueil des observations et des recherches qui ont été faites en Égypte pendant l'expédition de l'armée Française [1798-1801]. Histoire Naturelle, Paris, 2:1-218.

GERVAIS, P. 1855. Animaux nouveaux ou rares recueillis pendant l'expédition dans les parties centrales de l'Amérique du Sud, de Rio de Janeiro a Lima, et de Lima au Para; exécutée par ordre du gouvernement Français pendant les années 1843 a 1847, sous la direction du comte Francis de Castelnau. Mammifères. Paris, 1-116 pp. + 20 pls.

GOODWIN, G. G. 1942. A summary of recognizable species of Tonatia, with descriptions of two new species. J. Mamm., 23:204-209.

——. 1946. Mammals of Costa Rica. Bull. Amer. Mus. Nat. Hist., 87:271-473.

——. 1953. Catalogue of type specimens of Recent mammals in the American Museum of Natural History. Bull. Amer. Mus. Nat. Hist., 102:207-414.

——. 1959. Bats of the subgenus Natalus. Amer. Mus. Novit., 1977:1-22.

——. 1960. The status of Vespertilio auripendulus Shaw, 1800, and Molossus ater Geoffroy, 1805. Amer. Mus. Novit., 1994:1-6.

——. 1963. American bats of the genus Vampyressa, with the description of a new species. Amer. Mus. Novit., 2125:1-24.

GOODWIN, G. G., AND A. M. GREENHALL. 1961. A review of the bats of Trinidad and Tobago. Descriptions, rabies infection, and ecology. Bull. Amer. Mus. Nat. Hist., 122:187-302 + pls. 7-46.

GRAY, J. E. 1839. Descriptions of some Mammalia discovered in Cuba by W. S. MacLeay, Esq. Ann. Nat. Hist., 4:1-7 + 1 pl.

——. 1843. List of the specimens of Mammalia in the collection of the British Museum. British Museum, London, xxviii + 1-216.

HALL, E. R., AND J. W. BEE. 1960. The red fig-eating bat Stenoderma rufum found alive in the West Indies. Mammalia, 24:67-75.

HALL, E. R., AND W. W. DALQUEST. 1950. Pipistrellus cinnamomeus Miller 1902 referred to the genus Myotis. Univ. Kansas Publ., Mus. Nat. Hist., 1:581-590.

HALL, E. R., AND K. R. KELSON. 1959. The mammals of North America. Ronald Press, New York, 1:xxx + 1-546 + 79.

HALL, E. R., AND J. R. TAMSITT. 1968. A new subspecies of the red fig-eating bat from Puerto Rico. Life Sci. Occas. Papers, Royal Ontario Mus., 11:1-5.

HUSSON, A. M. 1962. The bats of Suriname. Zool. Verhand., Leiden, 58:1-282 + 30 pls.

JENTINK, F. A. 1887. Catalogue ostéologique des Mammifères. Mus. Hist. Nat. Pays-Bas, 9:1-366 + 12 pls.

——. 1888. Catalogue systématique des Mammifères (Rongeurs, Insectivores, Cheiroptères, Edentés et Marsupiaux). Mus. Hist. Nat. Pays-Bas, 12:1-280.

JOHNSON, D. H. 1952. A new name for the Jamaican bat Molossus fuliginosus Gray. Proc. Biol. Soc. Washington, 65:197-198.

JONES, J. K., JR., AND D. C. CARTER. 1976. Annotated checklist, with keys to subfamilies and genera. Pp. 7-38, in Biology of bats of the New World family Phyllostomatidae. Part I (R. J. Baker, J. K. Jones, Jr., and D. C. Carter, eds.). Spec. Publ. Mus., Texas Tech Univ., 10:1-218.

JONES, J. K., JR., H. H. GENOWAYS, AND R. J. BAKER. 1971. Morphological variation in *Stenoderma rufum*. J. Mamm., 52:244-247.

JONES, J. K., JR., P. SWANEPOEL, AND D. C. CARTER. 1977. Annotated checklist of the bats of Mexico and Central America. Occas. Papers Mus., Texas Tech Univ., 47:1-35.

LaVAL, R. K. 1973. A revision of the Neotropical bats of the genus *Myotis*. Sci. Bull. Nat. Hist. Mus., Los Angeles Co., 15:1-54.

MILLER, G. S., JR. 1897. Revision of the North American bats of the family Vespertilionidae. N. Amer. Fauna, 13:1-140 + 3 pls.

———. 1907. The families and genera of bats. Bull. U.S. Nat. Mus., 57:i-xvii + 1-282 + 14 pls.

———. 1913. Notes on the bats of the genus Molossus. Proc. U.S. Nat. Mus., 46:85-92.

MILLER, G. S., JR., AND G. M. ALLEN. 1928. The American bats of the genera Myotis and Pizonyx. Bull. U.S. Nat. Mus., 144:viii + 1-218.

PALLAS, P. S. 1766. Miscellanea zoologica quibus novae imprimis atque obscurae animalium species describunter et observationibus inconibusque illustratum. Hagae Comitum, xii + 1-224 pp. + i-xiv pls.

PELZELN, A. VON. 1883. Brasilische Säugethiere. Resultate von Johann Natterer's Reisen in den Jahren 1817 bis 1835. Verh. Kaiser Kön. zool.-bot. Ges., Wien, Beiheft zu Band 33, 1-140 pp.

PETERS, W. 1856. Uber die systematische Stellung der Gattung *Mormops* Leach und über die Classification der *Phyllostomata* so wie über eine neue Art der Gattung *Vampyrus*. Monatsb. Kön. preuss. Akad. Wiss. Berlin, 1856:409-415.

PINE, R. H. 1972. The bats of the genus *Carollia*. Tech. Monogr., Texas Agric. Exp. Sta., 8:1-125.

POOLE A. J., AND V. S. SCHANTZ. 1942. Catalog of the type specimens of mammals in the United States National Museum, including the Biological Surveys Collection. Bull. U.S. Nat. Mus., 178:xiv + 1-705.

RODE, P. 1938. Catalogue des types de mammifères du Muséum National d'Histoire Naturelle. I. Ordre des Primates: A.—scus—ordre des Simiens. Mus. Nat. Hist. Nat. Paris, pp. 1-52. [Reprinted from Bull. Mus. Hist. Nat., 2ᵉ sér., 10(3):202-251.]

———. 1941. Catalogue des types de mammifères du Muséum National d'Histoire Naturelle. II. Ordre des Chiroptères. Mus. Nat. Hist. Nat. Paris, pp. 71-105. [Reprinted from Bull. Mus. Nat. Hist. Nat., 2ᵉ sér., 13(4):227-252.]

ROUK, C. S., AND D. C. CARTER. 1972. A new species of Vampyrops (Chiroptera: Phyllostomatidae) from South America. Occas. Papers Mus., Texas Tech Univ., 1:1-7.

SANBORN, C. C. 1933. Bats of the genera Anoura and Lonchoglossa. Field Mus. Nat. Hist., Zool. Ser., 20:23-27.

———. 1937. American bats of the subfamily Emballonurinae. Field Mus. Nat. Hist., Zool. Ser., 20:321-354.

SCHINZ, H. R. 1821. Das Thierrich eingetheilt nach dem Bauder Thiere als Grundlage ihrer Naturgeschichte und der vergleichenden Anatomie von dem Herrn Ritter von Cuvier. I. Säugethiere und Vögel. J. G. Cotta'schen Buchhandlung, Stuttgart und Tübingen, i-xxxviii + 1-894.

SEBA, A. 1734. Locupletissimi rerum naturalium thesauri accurata descriptio, et iconibus artificiosissimis expressio, per universam physices historiam.... Amsterdam, 1:1-178 + 34 unnumbered pp. + 2 pls. + pls. i-cxi.

SHAMEL, H. H. 1931. Notes on the American bats of the genus Tadarida. Proc. U.S. Nat. Mus., 78:1-27.

SHERBORN, C. D., AND F. J. Griffin. 1934. On the dates of publication of the natural history portions of Alcide d'Orbigny's 'Voyage Amérique méridionale.' Ann. Mag. Nat. Hist., ser. 10, 13:130-134.

SIMPSON, G. G. 1945. The principles of classification and a classification of mammals. Bull. Amer. Mus. Nat. Hist., 85:xvi + 1-350.

SMITH, J. D. 1972. Systematics of the chiropteran family Mormoopidae. Misc. Publ. Mus. Nat. Hist., Univ. Kansas, 56:1-132.

———. 1977. On the nomenclatorial status of *Chilonycteris gymnonotus* Natterer, 1843. J. Mamm., 58:245-246.

TEMMINCK, C. J. 1826. Monographies de Mammalogie, ou description de quelques genres de mammifères, dont les espèces ont été observées dans les différens musées de l'Europe. Paris, 1:205-244 [vol. 1, 1824-1827, xxxii + 268 pp. + pls. i-xxv].

THOMAS, O. 1892. On the probable identity of certain specimens, formerly in the Lidth de Jeude collection, and now in the British Museum, with those figured by Albert Seba in his 'Thesaurus' of 1734. Proc. Zool. Soc. London, 22:309-318.

———. 1893. A preliminary list of the mammals of Trinidad. J. Trinidad Field Nat. Club, 1(7):158-168.

———. 1897. Descriptions of new bats and rodents from America. Ann. Mag. Nat. Hist., ser. 6, 20:544-553.

———. 1911. The mammals of the tenth edition of Linnaeus; an attempt to fix the types of the genera and the exact bases and localities of the species. Proc. Zool. Soc. London, 1911:120-158.

———. 1920. On mammals from the lower Amazons in the Goeldi Museum, Para. Ann. Mag. Nat. Hist., ser. 9, 6:266-283.

———. 1924. On a collection of mammals made by Mr. Latham Rutter in the Peruvian Amazons. Ann. Mag. Nat. Hist., ser. 9, 13:530-538.

TOMES, R. F. 1857. A monograph of the genus Lasiurus. Proc. Zool. Soc. London, 1857:34-46.

TORRE, L. DE LA. 1955. Bats from Guerrero, Jalisco and Oaxaca, Mexico. Fieldiana, 37:695-701 + 2 pls.

WAGNER, J. A. 1843. Diagnosen neuer Arten Brasilischer Handflüger. Arch. Naturgesch., 9(1):365-368.

———. 1845. Diagnosen einiger neuen Arten von Nagern und Handflüglern. Arch. Naturgesch., 1:145-149.

———. 1847. Beiträge zur Kenntniss de Säugthiere Amerika's. Aband. Kön. bayerischen Akad. Wiss., 5:119-208 + pls. ii-iv.

———. 1855. Die Säugthiere in Abbildungen nach de Natur mitt Beschreibungen von Dr. Johann Christian Daniel von Schreber. Supplementband, Leipzig, 5:i-viii + 1-810 + ix-xxvi.

WIED-NEUWIED, MAXIMILIAN PRINZ ZU. 1821. Reise nach Brasilien in den Jahren 1815 bis 1817. Frankfurt a.M., 2:v-xviii + 1-345 + pls. 15-22 + 1 map + vignettes i-viii.

———. 1822-1831. Abildungen zur Naturgeschichte Brasiliens. Recueil de planches coloriées d'animaux du Brésil. Weimar, 96 pls. [with text to each].

———. 1826. Beiträge zur Naturgeschichte von Brasilien. Verzeichniss der Amphibien, Säugthiere und Vögel, welche auf einer Reise Zwischen dem 13ten und dem 23sten Grade südlicher Breite im östlichen Brasilien beobachtet wurden. II. Abtheilung. Mammalia. Säugthiere. Weimar, 2:1-620 + pls. i-v.

WILSON, D. E. 1976. The subspecies of *Thyroptera discifera* (Lichtenstein and Peters). Proc. Biol. Soc. Washington, 89:305-311.

Addresses of authors: D. C. CARTER, *The Museum, Texas Tech University, Lubbock, 79409*; P. G. DOLAN, *The Museum and Department of Biological Sciences, Texas Tech University, Lubbock, 79409. Received 2 May, accepted 1 September 1977.*

APPENDIX 1.—External measurements. Genera and species are arranged in the order in which they appear in the text of this catalogue. Whenever possible, measurements of the forearm and metacarpals were taken from a flexed wing and include the carpals; skins preserved with the wings extended were measured in such a way as to approximate that of a flexed wing. Phalangial and tibial lengths are from proximal to distal epiphyses, length of ear is the greatest distance from notch to distal margin, and nose leaf was measured from the posterior base to the distal tip. Numbers in italic type are approximate measurements.

Species	Museum number	Length of forearm	Length of metacarpal III	Length of phalanx 1	Length of phalanx 2	Length of metacarpal IV	Length of phalanx 1	Length of metacarpal V	Length of phalanx 1	Length of tibia	Length of ear	Length of nose leaf (dry) from back	Length of calcar
EMBALLONURIDAE													
Rhynchonycteris													
Emballonura lineata	RNH 17642	39.6	40.5	12.5		41.5	8.7	31.3	9.4	13.7			
Proboscidea saxatilis	ZSM 22/50	38.8	38.6			32.5							
	ZMB 3085	41.6	40.8	12.8		34.6	8.9	32.3	9.4	15.6			
	RNH 17643	40.7	40.1	11.5		34.0	8.5	32.6	9.2	14.4			
Saccopteryx													
Emballonura insignis	NMW (not numbered)	44.9	43.2	12.8		38.8	8.0	36.1	9.4	18.0			
Saccopteryx canescens	BMNH 99.11.2.2	39.1	38.6	12.0		34.3	7.7	32.5	8.7	15.4			
Saccopteryx pumilia	BMNH 98.5.8.4	37.5	37.2	12.7		32.7	8.1	31.9	8.7	15.1			
Urocryptus bilineatus	RNH 17461	46.3	43.8	13.4		39.2	7.7	37.1	10.1	20.2			
Cormura													
Emballonura brevirostris	NMW (not numbered)	45.8	43.6	12.7		36.5	10.7	34.5	11.9	15.0			
	NMW (not numbered)	48.2	42.3	14.2		37.2	10.9	34.6	12.6	15.4			
Myropteryx pullus	ZMB 3360	44.3	40.7	13.3		34.6	10.0	32.7	11.3	14.9			
Peropteryx													
Emballonura macrotis	NMW (not numbered)	45.9	39.4	11.9		34.1		33.3	10.8	20.2			
Peronymus cyclops	BMNH 24.3.1.6	44.8	41.0	11.5		34.3	9.3	33.4	11.0	15.5			
Peropteryx kappleri	ZMB 3348	52.1	47.4	14.9		39.6	10.1	38.3	12.2	19.8			

APPENDIX 1.—*Continued.*

Centronycteris											
Centronycteris centralis	BMNH 0.7.11.3	44.8	47.9	18.4		38.5	9.5	39.0	9.8	20.4	
Vespertilio calcaratus	ZMB (not numbered)	46.2	47.4	17.9							15
Balantiopteryx											
Balantiopteryx plicata	ZMB 3361	41.3	36.7	10.8		30.2	8.7	30.1	9.8	16.0	
Diclidurus											
Diclidurus albus	ZMB 4478	64.0	63.2			49.0	13.7	38.2	17.1	19	
Diclidurus virgo	BMNH 98.10.9.3	66.1	65.6	9.4		50.0	13.8	38.4	18.4	24.5	
Cyttarops											
Cyttarops alecto	BMNH 12.11.4.5	46.1	45.2	9.3		39.3	9.2	33.0	13.1	19.1	
NOCTILIONIDAE											
Noctilio											
Dirias irex	BMNH 20.7.14.29	55.9	49.2	12.3		49.7	8.0	48.2	9.6	19	
Noctilio albiventer	ZSM 17		55.8	13.8	37.5	57.8	9.2	56.6	11.5		
	ZSM 102/128	52.9	40.4	11.7	24.6	42.2	7.6	41.5	7.8		
Noctilio rufus	ZSM 127	87.6	81.5	22.5	54.1	82.2	15.3	79.6	18.1		
Noctilio zaparo	MNCN 692	68.4	60.1	15.1		60.5	9.5				
MORMOOPIDAE											
Pteronotus											
Chilonycteris davyi fulvus	BMNH 93.2.5.24	43.5	40.6	8.9	14.6	35.6	8.5	33.3	9.8	16.0	
Chilonycteris gymnonotus	NMW (not numbered)	51	46.5	8.8	15.0	40.8	8.5	38.5	10.9	18.0	
Chilonycteris personata	NMW (not numbered)	47.2	42.5	8.7	12.8	37.0	7.5	34.7	9.6	17.0	
Chilonycteris rubiginosa	ZSM 45	64.2	55.2	12.1	17.3	52.7	12.4	51.5	12.2	24.0	
Mormoops											
Aello cuvieri	BMNH (not numbered)	48.5	41.3	9.8	18.6	39.2	11.0	32.4	14.9	19.5	
Mormoops megalophylla	ZMB 2826	54.5	50.7	10.2	23.5	46.5	12.9	36.5	16.8	20.5	
PHYLLOSTOMATIDAE											
PHYLLOSTOMATINAE											
Micronycteris											
Barticonycteris daviesi	BMNH 64.767	57.1	56.0	22.7	34.7	54.0	16.6	56.0	15.8	28.0	
Glyphonycteris sylvestris	BMNH 96.10.1.2	39.8	36.9	12.8	19.9	34.9	11.0	37.4	9.8	14.7	16
Phyllostoma elongata	BMNH 42.8.17.8		31.0	14.0	14.0	31.7	12.4		13.1		6

APPENDIX 1.—Continued.

Taxon	Specimen												
Phyllostoma scrobiculatum	NMW (not numbered)	34.2	29.4	13.1	14.3	30.4	9.9	31.2	9.9	12.7	20	5	
Schizostoma behnii	ZMB 5154	47.4	42.2	15.1	21.0	40.1	12.9	42.8	11.7	16.4	21	7	*8*
Schizostoma brachyote	MNHN 1876-1074	41.8	38.2	13.9	17.7	36.6	9.8	36.5	10.2	15.6	16	4	*13*
Schizostoma hirsutum	MNHN (not numbered)	41.4	35.3	16.7	17.2	36.6	14.2	39.0	14.2	17.6	20	6	*12*
Tonatia													
Lophostoma brasiliense	BMNH 49.11.7.14	38.8	31.1	13.2	14.3	32.1	12.5	32.5	12.9	17.5	18.6	6.5	*13*
Phyllostoma amblyotis	NMW (not numbered)	54.5	44.0	18.9	21.1	45.8	17.3	49.0	16.7	23.3	33	6	
Phyllostoma childreni	BMNH 8a	51.8	43.8	16.7	21.0	44.7	12.7	44.7	14.8	19.0		8	*14*
Tonatia laephotis	BMNH 10.5.4.5	55.5	46.9	19.1	20.8	47.0	18.4	50.6	17.7	26.8		12	*17*
Vampyrus bidens	ZSM (not numbered)	57.0	49.3	19.9		50.0	15.5	51.6	16.2		20		
Mimon													
Anthorhina peruana	BMNH 23.10.16.12	48.2	45.0	13.5	25.0	45.0	10.4	44.2	9.8	23.2	13.5	13.0	
Anthorhina picata	BMNH 3.9.5.26	48.9	47.9	15.0	16.2	47.7	12.3	48.6	11.8	20.0		15	*23*
Phyllostomus													
Phyllostoma discolor	ZSM 133	61.6	54.0	13.9	25.8	53.5	10.7	53.0	9.8				
Phyllostoma elongatum	MNHN A.2	65.9	58.8	16.5	28.7	57.6	13.5	58.8	13.5	25.1	28	12	*21*
Phyllostoma latifolium	BMNH 1.6.4.43	58.7	54.4	16.4	29.5	53.2	13.8	54.2	12.7	20.6	22	11	
Phylloderma													
Guandira cayanensis	BMNH 42.10.25.2	73.0	65.4	20.8	30.2	63.6	15.0	67.6	14.9	27.5	20	8	
Phylloderma stenops	RNH 16843	73.3	68.1	24.3	32.9	66.4	17.6	69.2	16.5	*31*		7	
Trachops													
Trachops cirrhosus ehrhardti	SMF 11716	59.3	50.3	20.6	31.0	52.7	17.6	54.5	16.7	23.3	31		
	SMF 11717	60.0	52.3	20.7	30.4	53.3	17.6	55.6	16.2	23.7			
	SMF 11718	56.2	47.4	19.4	28.4	49.8	16.2	51.7	15.2	22.7			
Chrotopterus													
Chrotopterus auritus australis	BMNH 1.3.11.1	82.5	63.5	32.9	34.7	67.7	21.2	73.3	24.7	35.7	39	10	
Chrotopterus auritus guianae	BMNH 4.5.7.20	81.0	60.9	31.0	36.1	65.5	24.5	73.1	24.1	36.4	38	11	
Vampyrus auritus	ZMB 3755	84.6	65.1	31.7	35.6	68.6	24.9	74.7	25.3	34.3	47	13	
GLOSSOPHAGINAE													
Glossophaga													
Monophyllus leachii	BMNH 42.8.17.17	36.4	36.0	12.0	15.4	32.8	9.0	32.0	7.7	12.5			*3.4*

APPENDIX 1.—Continued.

Taxon	Specimen												
Lonchophylla													
Lonchophylla mordax	BMNH 3.9.5.34	34.5	32.3	11.7	15.8	31.0	8.9	29.9	8.2	13.1			6.6
Lionycteris													
Lionycteris spurrelli	BMNH 13.8.10.1	32.8	31.7	9.7	16.7	29.3	7.5	26.4	7.5	12.1		4.2	4.0
Anoura													
Anoura geoffroyi	BMNH 11a	42.9	43.9	13.5	22.4	42.5	9.6	36.5		14.7			3.2
Glossonycteris lasiopyga	ZMB 3564	45.6		14.5	22.6		10.5	36.9	8.9				
	ZMB 3565	43.0	42.3	14.2	21.3	40.9	10.7	34.4	8.2	14.2			
Glossophaga caudifer	MNHN 937	36.8	37.1	11.6	18.6	34.8	9.5	31.3	7.0	12.8			
Lonchoglossa wiedi aequatoris	NR 6	35.5	37.9	12.2	19.6	35.7	9.2	30.8	8.0				
	NR 8		38.1	11.8	20.0	35.6	8.6	30.9	7.1	11.4			
Scleronycteris													
Scleronycteris ega	BMNH 7.1.1.671	34.8	34.5	14.7	17.7	33.0	9.2	30.9	8.5				7.0
Hylonycteris													
Hylonycteris underwoodi	BMNH 3.2.1.5	34.7	35.3	13.3	18.1	33.0	8.6	29.3	7.6	11.6		4.9	6.0
Platalina													
Platalina genovensium	BMNH 27.11.19.38	45.9	45.9	18.4		43.5	13.2	42.4	12.5	18.3			8
Choeroniscus													
Choeronycteris godmani	BMNH 79.12.24.1	33.4	34.2	12.7	16.8	32.4	8.8	29.7	8.6	11.8	11	3	5
Choeronycteris inca	BMNH 12.9.5.2	33.2	34.7	11.5	16.4	31.7	8.6	30.2	7.5	11.8			6
Choeronycteris minor	SMNS 441	34.6	35.6	12.0	17.7	33.0	8.9	31.5	8.1	11.1	12	3	6
CAROLLIINAE													
Carollia													
Carollia azteca	ZMB 2647	43	40	17.6	20.9	38.9	13.7	41.2	12.8	16.1	12.9	7.8	
Carollia verrucata	BMNH 106.a	38.6	37.6	15.7	19.3	35.4		37.1	11.2	16.0			
Glossophaga amplexicauda	MNHN A. 291		35.2	15.3	18.1	33.8	11.8	34.8	10.5	14.1			
	MNHN 935	40.6	37.9	15.8	20.5	36.8	12.7	39.4	10.8	15.7			
	MNHN 936	42.4	40.7	17.0	22.3	37.8	13.5	39.6	11.8	15.8			
Phyllostoma bicolor	ZSM 126		42.8	17.7	24.0	40.2	14.5	41.0	13.0	16.2	17		
Phyllostoma calcaratum	ZSM 141	42.3	39.6	16.1	21.2	37.9	12.8	40.5	12.0				
Rhinops minor	BMNH 49.10.15.13	36.0	30.3	12.2	16.2	29.5				14.2	14	5	
Vespertilio perspicillatus	BMNH 67.4.12.597	39.2	36.0	14.7	18.0	35.3	12.1	37.1	11.9	15.9			6.2

APPENDIX 1.—Continued.

Taxon	Catalog													
Rhinophylla														
Rhinophylla pumilio	ZMB 3060	33.8	31.1	13.9	18.4	31.7	11.9	32.0	9.7	11.7	14	7.1	3.9
Rhynophylla cumilis	SMNS 289 (1)	34.5	34.9	18.6	13.7					10.7			
	SMNS 289 (2)	33.7	30.9	17.7	13.7					10.7			
	ZMB 3346	34.7	34.7	14.6	19.6	34.9	12.2	35.5	9.2	11.4			
STENODERMINAE														
Sturnira														
Corvira bidens	BMNH 15.7.11.7		49.2	16.4	21.5	42.3	14.7	44.2	12.2	15.5			
Phyllostoma albescens	ZSM 129	43.3	39.2	15.6	18.6	38.4	12.3	40.2	10.4				
	NMW (not numbered)	41.3	40.5	14.9	18.6	40.1	11.8	41.7	8.8	15.1			
Phyllostoma excisum	ZSM 137	43.8	41.4	15.2	19.6	41.6	12.6	42.1	9.8	15.2			
Phyllostoma fumarium	ZSM 58	45.2	39.4	15.3	19.5	40.2	12.4	41.6	9.2				
Phyllostoma lilium	MNHN (not numbered)	43.2	42.2	15.5	20.5	41.9	12.9	43.8	9.6	15.5			
Phyllostoma oporophilum	ZSM 145	48.5	45.6	17.6	22.7	44.6	13.8	47.3	9.8	16.0			
Sturnira spectrum	BMNH 42.12.2.4		41.0	14.9	18.4	40.6					15		
Uroderma														
Uroderma bilobatum	ZMB 409	44.5	43.7	13.8	23.6	42.6	12.3	42.4	8.5	15.1			
	ZMB 410	44.6	45.1	14.5	24.1	41.7	12.4	43.0	9.2	16.6			
	ZMB 411	41.1	41.5	15.4	21.3	39.6	12.7	41.0	9.9	14.7			
Vampyrops														
Artibeus vittatus	ZMB 568	59.5	57.5	22.2	31.3	55.2	19.4	56.6	14.4	19.0			
Phyllostoma lineatum	MNHN 953	46.9	45.5	17.8	24.5	44.5	14.5	47.4	10.6	16.6		6	4.2
Vampyrops dorsalis	BMNH 99.12.5.1	48.2	47.1	20.2	25.8	46.4	16.1	47.2	12.3	18.0		8.2	
Vampyrops helleri	ZMB 3276	38.5	41.4	12.6	22.9	39.8	11.1	40.3	8.6	14.5			
Vampyrops lineatus sacrillus	BMNH 23.12.12.9	47.2	46.7	16.8	23.6	45.6	9.8	47.5	10.0	14.8		8.0	4.5
Vampyrops oratus	BMNH 14.7.27.1	46.9	44.5	17.2	24.4	43.7	14.8	45.3	11.3	17.2		10.0	4.8
Vampyrops recifinus	BMNH 81.2.16.4	40.2	39.2	13.9	24.1	38.6	12.3	39.6	9.3	13.9		11	3
Vampyrops zarhinus incarum	BMNH 12.1.15.1	36.0	38.0	12.8	21.9	36.9	10.9	37.7	8.6	11.2		10	4
Vampyrodes														
Vampyrodes ornatus	BMNH 24.3.1.63	53.9	52.9			52.1		53.5		17.5			
Vampyrops caracciolae	BMNH 89.6.10.2	49.7	47.8	17.2	27.5					15.9			

APPENDIX 1.—*Continued.*

Taxon	Catalog No.											
Vampyressa												
Chiroderma bidens	BMNH 69.3.31.12	35.6	37.6	12.4	20.7	34.8	10.6	35.3	7.6	12.5		5
Phyllostoma pusillum	ZSM 1843/2	33.2	31.5	13.6	16.0	31.3	10.6	31.7	8.9	11.6		4
Vampyressa melissa	BMNH 26.5.3.4	37.2	35.6	15.3	20.4	35.0	13.6	36.7	11.1	13.1		
Vampyressa nymphaea	BMNH 9.7.17.40	34.7	34.7	12.2	19.2	33.9	10.7	34.7	7.8	11.9		
Vampyressa venilla	BMNH 24.3.1.73	30.5	29.8			28.5		30.0		11.3		
Chiroderma												
Chiroderma doriae	BMNH 44.9.2.6		52.8	20.5	28.5	51.7	17.1	53.4	12.9	18		6
Chiroderma salvini	BMNH 68.8.16.2	50.6	49.2	18.9	27.7	49.3	15.6	50.0	11.6	16.5		
Chiroderma villosum	ZMB 408	46.9	48.5	17.3	25.6	44.3	15.7	45.6	11.3	15.6		
Ectophylla												
Mesophylla macconnelli	BMNH 1.6.4.64	30.0	28.3	12.2	14.6	27.6		28.8	8.0	9.4		6.2
Artibeus												
Artibeus carpolegus	BMNH 47.12.27.13	58.6	56.2	16.5	27.7	54.9	14.5	56.4	11.5			
Artibeus cinereus bogatensis	BMNH 99.11.4.35	43.5	41.3	15.8	21.4	39.6	13.1	41.5	10.7	14.3		
Artibeus concolor	ZMB 2617	46.8	46.7	15.7	26.5	46.0	13.7	46.2	11.2	15.6	17	6
Artibeus fallax	RNH 13083	65.0	61.0	18.5	30.6	60.0	15.4	61.2	10.7	21.0		
	ZMB 566	66.7	62.1	19.6	33.6	62.3	17.2	63.7	13.4	24.1	23	8
Artibeus glaucus	BMNH 94.8.6.13	42.8	40.0	15.8	23.3							
Artibeus jamaicensis	BMNH (not numbered)		56.5	15.6	26.6	56.3	24.3	56.6	10.6			
Artibeus jamaicensis aequatorialis	BMNH 0.2.9.13	69.8	57.5	17.2	31.0	56.4	15.2	56.0	12.0	20.5		
Artibeus nanus	BMNH 89.1.30.5	35.5	33.3	12.2	17.5	32.7	10.4	35.5	8.6	12.6		
Artibeus planirostris trinitatis	BMNH 97.6.7.1	57.1	53.2	18.8	28.9	54.0	15.9	53.9	12.1	19.8		
Artibeus pumilio	BMNH 24.3.1.52	37.5	35.6	13.3	18.7	35.1	11.3	36.0	9.0	13.1		5
Artibeus (Dermanura) quadrivittatus	RNH 13114	42.2	37.6	14.5	19.7	37.7	12.0	39.2	10.2			
Artibeus (Dermanura) rosenbergi	BMNH 97.11.7.76	36.7	36.8	13.4	19.3	35.4	11.2	35.6	8.0	12.2		3
Artibeus turpis	BMNH 88.8.8.29	39.2	38.3	13.6	20.7	37.8	11.1	38.1	9.8	12.2		
Artibeus watsoni	BMNH 0.7.11.19	38.5	37.2	14.1	21.4	37.2	11.8	38.4	8.9	14.0		
Madatacus lewisii	BMNH b 4.b	51.4										
Enchisthenes												
Artibeus hartii	BMNH 92.9.7.8	37.2	35.7	13.1	16.1							

APPENDIX 1.—Continued.

Taxon	Specimen										
Stenoderma											
Stenoderma rufa	MNHN 934		46.3	11.1	23.6	44.2	12.5	46.5	10.6		3
Pygoderma											
Phyllostoma bilabiatum	ZSM 114	41.7	41.5	17.7	28.8	42.8	12.7	44.0	11.3	18.2	17
Stenoderma (Pygoderma) microdon	ZMB 2713	38.1	38.5	16.8	26.2	38.9	11.6	40.6	10.4	15.9	8
Sphaeronycteris											
Sphaeronycteris toxophyllum	ZMB 598	40.2	41.3	14.7	22.1	40.7	11.3	41.4	11.2	16.7	
Brachyphylla											
Brachyphylla cavernarum	BMNH (not numbered)	65.8	75.4	25.7	11.1	55.0	15.7	57.4	14.7	29.7	
DESMODONTINAE											
Desmodus											
Desmodus dorbignyi	BMNH (not numbered)	59.3	54.6	10.2	18.5	54.8	9.0	53.7	9.2		
Desmodus murinus	ZSM 57	60.6	57.7	11.8	19.0	56.2	10.3	53.8	10.0		
	ZSM 59	60.3	52.7	10.6	17.9	52.8	9.4	51.1	9.6		
Edostoma cinerea	MNHN 958D	63.6	59.0	11.1	18.9	57.6	10.2	57.6	9.6	15.3	
	MNHN 958G	60.6	55.5	9.9	19.5	55.7	9.4	54.8	10.2	26.8	
Diaemus											
Desmodus youngii	RNH 12088	49.4	50.0	9.4	25.2	49.2	9.0	47.6	8.6	20.1	15
Diaemus youngi cypselinus	BMNH 28.7.21.64	56.9	55.7	11.1	30.2	55.5				22	
Diphylla											
Diphylla centralis	BMNH 3.3.3.3	54.9	55.6	11.2	27.8	56.0	10.8	54.1	12.8		
NATALIDAE											
Natalus											
Phodotes tumidirostris continentis	BMNH 11.5.25.13	38.1	34.8	15.0		35.8	9.2	35.4	9.2	18.3	
VESPERTILIONIDAE											
Myotis											
Myotis dinellii	BMNH 0.7.9.4	38.8	35.8	10.4	10.6	34.8	9.0	34.2	9.0	16.1	13
Myotis thysanodes aztecus	BMNH 58.6.2.3	43.9	40.7	14.9	11.2	40.1	11.4	40.2	10.3	15.0	
Vespertilio arsinoe	RNH 17635	35.6	31.3	12.2	10	31.1	8.8	30.7	8.7	12.1	9
Vespertilio carbonarius	ZSM 124	35.9	33.2	11.6		31.9	8.9	30.4	7.8	14.3	
Vespertilio leucogaster	RNH 17622	36.5	30.8	10.3	8.7	31.5	9.2	30.6	8.0	14.1	

APPENDIX 1.—*Continued.*

Taxon	Catalog										
Vespertilio levis	MNHN 864	41.4	37.0	11.2	11.7	35.6	9.3	35.0	9.3	16.1	14
Vespertilio nubilus	ZSM 121	39.1	36.8	11.8		35.6	9.6	35.2	8.9		
Vespertilio parvulus	RNH 17621	33.2	29.6	10.2	9.7	29.6	8.1	28.6	7.4	13.4	
Vespertilio (*Leuconoe*) *pilosus*	MNHN (not numbered)	54.4	51.3	17.3	13.3	49.7	14.2	49.3	13.3	19.0	
Vespertilio polythrix	MNHN 843	35.2	33.2	10.5	15.0	32.1	8.2	31.2	7.7	14.6	
	MNHN 842	39.2	36.5	11.4		36.6	9.2			15.7	
Vespertilio splendidus	ZSM 142	39.7	37.6	12.5		37.4	9.2	35.7	8.0	14.9	
Pipistrellus											
Vespertilio isidori	MNHN 865	35.3	33.7	12.2		32.9	11.0	31.8	9.3		
Eptesicus											
Eptesicus argentinus	BMNH 98.3.4.6	45.2	42.3	15.0	12.2	42.5	13.4	40.3	9.6	14.5	
Eptesicus chiriquinus	BMNH 3.3.3.1	46.1	43.3	16.3	14.7	42.1	15.3	40.3	11.1	17.8	
Vesperugo (*Vesperus*) *dorianus*	BMNH 86.11.3.13	38.2	36.7	14.1	11.2	35.4	12.4	12.4	9.0	15.6	
Eptesicus fuscus pelliceus	BMNH 98.7.1.28	53.7	50.4	19.4	16.1	49.4	17.2	48.6	12.8	20.7	
Eptesicus montosus	BMNH 2.1.1.1	42.8	40.0	13.8	11.5	39.2	12.2	37.8	9.7	15.9	
Eptesicus punicus	BMNH 99.8.1.1	34.9	33.7	11.8	11.4	32.5	10.2	31.0	7.6	14.2	
Scotophilus cubensis	BMNH 103a	46.0	43.7	15.6	14.9	43.5	15.6	40.8	10.6	7.8	
Scotophilus macleayi	BMNH 104a	47.7	45.0	16.8	14.7	44.7	14.7	42.2	11.3	18.9	
Vespertilio arctoideus	ZSM 144	42.8	39.6	15.6	12.2	39.2	13.2	38.1	10.5		
Vespertilio fuscus peninsulae	BMNH 95.34.10.14	43.3	40.7	15.8	11.9	39.8	13.8	38.8	9.8	16.7	
Vespertilio hilarii	ZMB 3912	41.6	38.4	14.2	12.2	37.8	13.2	36.6	10.9	14.9	
Vesperus melanopterus	RNH 12092	40.4	38.8	14.8	12.5	38.0	13.0	36.3	9.4	14.3	13
Nycticeius											
Vespertilio aenobarbus	RNH 17623	29.0	28.2	11.0	10.1	26.6	9.4	27.1	7.5	11.3	
Rhogeessa											
Rhogeessa io	BMNH 94.9.25.1	28.1	27.3	10.6	9.2	27.0	8.7	27.5	7.2	10.4	11
Rhogeessa velilla	BMNH 99.8.1.5	27.9	27.5	9.9	7.9	26.8		27.2		8.8	11
Lasiurus											
Atalapha egregia	ZMB 3762	49.3	59.6	20.6	22.7	52.6	12.6	45.5	9.3	20.8	15
Atalapha frantzii	ZMB 2707	40.3	45.9	17.1	18.3	42.1	11.7	37.7	7.7	17.2	
	ZMB 2707	39.2	43.2	16.2	18.7	39.2	10.6	36.2	7.3	18.8	
	ZMB 3451	40.9	46.0	16.7	18.1	41.6	12.2	37.7	8.1	19.2	

APPENDIX 1.—*Continued.*

Atalapha pallescens	ZMB 595	55.8	62.8	19.6	25.0	56.1	12.0	48.6	8.9	20.5
Dasypterus ega argentinus	BMNH 98.3.4.9	44.2	50.2	16.5		46.8	12.2	42.3	7.5	
Dasypterus ega fuscatus	BMNH 99.9.6.5	48.1	52.9	18.5		48.6	12.7	42.4	7.9	18.9
Dasypterus ega panamensis	BMNH 0.7.11.1	46.4	55.3	17.3		49.6	11.4	43.3	7.2	18.2
Dasypterus ega xanthinus	BMNH 98.3.1.14	47.3	51.0	18.1		47.8	12.4	40.7	8.2	18.0
Lasiurus caudatus	BMNH 44.10.9.7	44.4	49.0	16.2		45.4	11.7			15.5
MOLOSSIDAE										
Molossops										
Dysopes abrasus	RNH 17374	42.0	43.6	18.6	15.9	41.7	15.9	27.2	10.7	
Dysopes temminckii	ZMB 5458	31.6	33.0						7.7	
Molossops mastivus	BMNH 10.11.10.3		50.6	20.6	17.7	47.9	16.9	28.6	12.3	
Molossops temminckii sylvia	BMNH 98.3.4.21	29.9	33.0							
Molossus cerastes	BMNH 1.8.1.13	45.0	48.3	21.9	18.8	45.9	18.3	29.7	12.7	
Molossus (Molossops) planirostris	ZMB 2513	32.5	34.6	14.3	13.7	33.5	12.4	21.6	9.2	8.5
Tadarida										
Dysopes auritus	NMW (not numbered)	63.3	60.8	24.0	20.5	57.2	20.2	31.8	20.5	
Dysopes gracilis	ZSM 135	43.2	42.3	18.7	15.6	40.6	15.6	25.2	15.5	11.6
	ZMB 2467	43.2	42.6	18.5	16.6	41.7	16.0	25.9	13.6	11.7
Dysopes multispinosus	ZMB 2614	42.7	44.1	16.0	15.0	42.3	13.3	26.3	13.6	11.3
	ZMB 2614	41.1	43.2	15.8	15.5	42.2	13.5	25.8	13.7	10.8
Dysopes naso	ZMB 2464	44.7	42.5	17.2	15.6	41.2	14.1	26.6	13.9	12.4
	ZSM 130		45.4	17.7	16.6	43.7	14.0	27.7	14.0	11.7
	ZSM 134	44.8	45.1	16.6	14.5	43.2	13.8	27.5	14.1	12.0
Dysopes nasutus	RNH 17575	44.3	43.2	15.7	13.0	42.4	13.4	26.0	13.0	12.6
	RNH 17576	44.9	44.5	16.3	15.7	43.3	13.9	27.0	13.6	12.3
Molossus mexicanus	ZMB 2589	41.5	42.6	15.7	15.2	40.8	13.2	25.2	12.7	11.5
Molossus rugosus	MNHN 795J	44.6	45.2	16.0		43.5	13.6	27.1	13.4	12.8
	MNHN 795K		46.2	17.7	17.4	43.8	14.7	27.1	13.9	12.2
Nyctinomus brasiliensis	MNHN 800	46.7	45.9	17.1	16.8	45.0	14.7	27.8	14.2	13.0
	MNHN 801	43.3	45.0	17.7		43.5	15.1	27.6	12.4	12.3
	MNHN 802	45.8	45.5	15.5		43.6	13.6	28.3	13.6	12.3
	MNHN 803	45.7	43.5	16.5		43.8	13.5	26.9	14.1	12.7

APPENDIX 1.—Continued.

Taxon	Specimen										
Nyctinomus macrotis	BMNH (not numbered)	58.3	57.1	22.3	20.2	54.1	18.5	27.9	17.6	17.1	
Nyctinomus musculus	ZMB 2457	39.9	40.3	14.3		39.2	12.2	23.1	11.7	10.5	
	ZMB 2457	38.3	40.2	13.4		38.6	12.2	23.2	11.5	9.4	
	ZMB 2457	39.6	40.6	14.7		39.5	12.2	23.4	12.0	9.5	
	ZMB 2457	40.0	39.9	14.6		38.8	12.8	23.9	12.3	9.9	
	ZMB 2482	39.7	40.2	14.2		38.2	11.6	23.3	11.6	10.6	
Eumops											
Dysopes glaucinus	NMW (not numbered)	60.2	60.3	16.0	23.2	58.6	20.7	34.2	17.8		
Dysopes longimanus	ZSM 56	60.1	58.3	25.5	22.9	57.1	20.8	33.7	18.1		17
	ZSM 60	62.7	60.5	27.0	24.2	58.7	22.0	33.3	17.7		
Dysopes (Molossus) gigas	ZMB 2474	74.9	73.8	29.8	27.6	71.3	25.5	40.3	22.0		
Eumops dabbenei	BMNH 14.4.4.8	77.5	77.7	31.8	30.3	76.4	26.5	44.3	23.2	24.3	
Eumops delticus	BMNH 23.8.9.7	46.9	48.8	19.2	18.2	46.9	16.8	26.9	15.2		
Molossus ferox	ZMB 2587	61.5	62.7	26.5	24.7	60.3	20.8	34.2	18.5		
	ZMB 2865	62.1	63.2	25.5	24.8	61.6	20.9	33.6	18.1		
	ZMB 2871	61.2	61.4	25.4	23.0	59.2	20.8	34.1	19.1		
	ZMB 2981	62.1	62.4	26.0	24.4	60.7	21.1	34.6	18.5		
Molossus maurus	BMNH 1.6.4.34	53.4	54.7	24.1	21.8	51.9	20.3	30.7	16.2	17.3	
Promops nanus	BMNH 0.7.11.99	38.5	39.8	15.8	14.9	38.7	13.2	22.7	12.2	10.8	
Promops trumbulli	BMNH 99.11.2.1	72.7	71.7	28.0		70.6		40.0	21.3	22.5	
Promops											
Molossus fosteri	BMNH 1.8.1.17	48.1	51.3	20.8	17.4	50.0	18.0	31.0	12.6	15.6	
Molossus nasutus	ZSM 136	51.7	51.8	21.6	18.0	49.9	18.7	32.9	13.2		
Promops ancilla	BMNH 6.5.8.4	49.6	54.2	21.9	19.2	51.7	18.5	33.5	12.7	17.6	
Promops centralis	BMNH 94.2.5.4	53.4	56.3	24.7	21.0			34.2		18.5	
Promops davisoni	BMNH 21.5.21.1	51.0	55.1			52.8		34.4	13.6	17.7	
Promops occultus	BMNH 2.11.7.24	52.0	56.1	23.4	19.4	54.4	21.0	36.1	15.5	17.0	
Molossus											
Dysopes alecto	RNH 13023	54.7	52.2	24.5	21.6	50.6	20.6	32.8	13.1		
Molossus burnesi	BMNH 5.1.8.7	33.8	36.0	15.6	13.5	34.7	13.5	21.7	7.5	11.6	
Molossus fuliginosus	BMNH 28a	36.7	40.1	17.6		38.8	15.0	26.1	10.8	12.0	
Molossus longicaudatus	MNHN 792	36.9	39.6			38.5		25.6	11.2		

APPENDIX 1.—*Continued.*

Molossus obscurus	MNHN A.419/225	39.9	41.2	18.4	15.7	39.8	15.6	26.1	10.6	12.9	13
	MNHN A.419/225a	41.0	42.4			40.9		26.8		13.1	
Molossus obscurus currentium	BMNH 98.3.4.28	41.7	43.3	19.0		42.3		28.9	11.2	13.5	
Molossus rufus	MNHN A.428/224	50.9	51.4	23.5	20.2	49.6	20.1	31.5	13.6	17.8	16
	MNHN A.428/224a	49.4	52.4	23.5	20.9	51.1	20.2	31.6	13.6	17.5	15

APPENDIX 2.—Skull measurements. Genera and species are arranged in the order in which they appear in the text of this catalogue. Greatest length of skull, condylobasal length, and palatal length include the incisors; greatest length of mandible and length of mandibular toothrow also include the incisors when normally they are components of the dental arcade. Depth of the braincase was measured with the aide of a microscope slide, against which the base of the cranium was placed (this measurement was not taken for skulls lacking one or both of the tympana). Length of maxillary toothrow, greatest breadth across molars, greatest breadth across canines, and length of mandibular toothrow are the maximal measurements possible for those variates.

Species	Museum number	Greatest length of skull	Condylobasal length	Palatal length	Zygomatic breadth	Mastoidal breadth	Breadth of braincase	Depth of braincase	Least postorbital breadth	Length of maxillary toothrow	Greatest breadth across upper molars	Greatest breadth across upper canines	Greatest length of mandible	Length of mandibular toothrow
EMBALLONURIDAE														
Rhynchonycteris														
Emballonura lineata	RNH 17642	12.1			7.0	6.5	6.0	6.3	2.4	4.6	4.8	3.2		4.4
Saccopteryx														
Saccopteryx bilineata centralis	BMNH 88.8.8.20	16.4			10.1	8.3	7.8	8.2	2.5	6.4	6.6	3.9	10.8	6.7
Saccopteryx canescens	BMNH 99.11.2.2	13.1	11.5			6.9	6.5		2.3	4.8	5.4	2.6		5.1
Saccopteryx gymnura	BMNH 75.10.22.2	12.9	11.4		7.8	6.6	6.2	6.3	2.2	4.6	5.0	2.9	8.2	4.6
Saccopteryx pumilia	BMNH 98.5.8.4	12.6	11.2			6.6	6.3		1.9	4.8		2.8		4.9
Urocryptus bilineatus	RNH 17461									7.3	7.3	4.2		7.6
Cormura														
Emballonura brevirostris	NMW (not numbered)	16.5				8.8	8.0	7.8	3.0	6.6	6.7	3.6		6.5
Peropteryx														
Peronymus cyclops	BMNH 24.3.1.6	16.3							3.2	6.5	7.6	4.7		6.7
Centronycteris														
Centronycteris centralis	BMNH 0.7.11.3								3.1	6.0	6.6	3.6		6.1
Balantiopteryx														
Balantiopteryx io	BMNH 86.9.3.1	12.9	11.5			7.6	6.6	6.4	3.3	4.8	5.9	3.7		4.7

APPENDIX 2.—*Continued.*

Taxon	Specimen													
Diclidurus														
Diclidurus virgo	BMNH 98.10.9.3	19.0	18.0			10.0	9.2	9.7	5.6	8.1	8.7	4.6		8.9
Cyttarops														
Cyttarops alecto	BMNH 12.11.4.5	13.5	12.8	8.0		7.3	6.7	7.0	3.5	5.4	6.0	3.0		5.7
NOCTILIONIDAE														
Noctilio														
Dirias irex	BMNH 20.7.14.29	18.6	16.5	8.2		12.8	10.8	9.9	5.5	6.9	8.2	6.2	11.8	7.4
Noctilio rufus	ZSM 127	29.2												
MORMOOPIDAE														
Pteronotus														
Chilonycteris davyi fulvus	BMNH 93.2.5.24	15.4	14.6	8.2		8.4			3.5	6.3	5.6	4.8		
Chilonycteris gymnonotus	NMW (not numbered)	17.5								7.0	6.8	5.6		
Chilonycteris personata	NMW (not numbered)	16.0												
Chilonycteris psilotis	BMNH 50.8.29.3	15.4	13.8	6.6		8.8	7.8	7.6	3.6				10.0	6.2
Pteronotus davyi	BMNH 9.1.4.74	15.8	15.0	7.5		8.6	7.8	8.1	3.7	6.6	6.0	4.8	11.0	7.0
Mormoops														
Aello cuvieri	BMNH (not numbered)	13.8	13.8			7.3	7.1	8.5	4.3	7.5	6.0	4.2	11.8	8.2
Mormops megalophylla	ZMB 2826	14.7	15.0	8.9		8.3	8.7	9.7		8.2	6.8			8.9
PHYLLOSTOMATIDAE														
PHYLLOSTOMATINAE														
Micronycteris														
Barticonycteris daviesi	BMNH 64.767	27.3	26.0	13.8	13.4	11.4	10.9		6.5	11.1	9.3	5.2	19.5	12.0
Glyphonycteris sylvestris	BMNH 94.10.1.2	20.2	18.2	9.4		9.1	8.4	9.2	4.6	7.9	6.9	3.4	12.7	8.5
Phyllophora megalotis	BMNH (not numbered)	17.9	15.8	8.1		7.6	7.1	8.4	3.7	6.7	5.8	3.2	11.1	7.3
Phyllostoma elongata	BMNH 42.8.17.8									6.8		3.1	11.4	7.3
Phyllostoma scrobiculatum	NMW (not numbered)	18.4												
Schizostoma behnii	ZMB 5154	23.5	21.2			10.0	9.2	9.8	5.1	9.3	7.6	4.0	15.1	9.6
Schizostoma brachyote	MNHN 1876-1074	21.7	19.5		10.8	10.0	8.7		5.1	8.5	7.0	4.2		9.0
Schizostoma hirsutum	MNHN (not numbered)	23.3	20.5			10.3	8.8		4.8	9.1	7.1	4.0	4.8	9.8
Lonchorhina														
Lonchorhina aurita	BMNH 9.1.4.67	20.6	19.1	9.6		10.3	8.8	8.7	4.8				12.5	7.5

APPENDIX 2.—Continued.

Taxon	Catalog no.													
Tonatia														
Lophostoma brasiliense	BMNH 49.11.7.14								3.6	7.5	6.4	4.4	12.6	8.2
Phyllostoma amblyotis	NMW (not numbered)	27.2												
Phyllostoma childreni	BMNH 8a			11.9	13.0									10.2
Tonatia laephotis	BMNH 10.5.4.5	28.8	24.9	13.4	14.5	13.3	11.0	13.5	4.3	10.4	9.2	6.3	16.2	11.5
Vampyrus bidens	ZSM (not numbered)	28.6	24.8	12.9	13.8	13.3	11.4	13.5	6.0	10.0	8.8	6.2	17.9	11.2
Mimon														
Anthorhina peruana	BMNH 23.10.16.12	21.9	19.0	9.8	11.9	11.1	8.4	10.4	3.7	8.0	8.3	5.3	13.4	8.7
Anthorhina picata	BMNH 3.9.5.26	22.0	19.2	9.8	12.5	12.2	8.7	10.6	4.0	7.7	8.1	5.2	13.5	8.4
Phyllostomus														
Phyllostoma discolor	ZSM 133			14.3			10.8	13.5	5.9	10.6	11.0	7.3	18.8	12.0
Phyllostoma elongatum	MNHN A.2	29.7	25.9									7.5		11.5
Phyllostoma latifolium	BMNH 1.6.4.43	28.2	25.0	13.2	15.8	13.7	10.6	11.8	5.0	10.4	11.0	6.9	17.9	11.1
Phylloderma														
Guandira cayanensis	BMNH 42.10.25.2							10.2				6.0		
Phylloderma stenops	RNH 16843	30.0		13.8	14.5	13.5	12.6	8.6		9.8	9.3	5.9		10.5
Trachops														
Trachops cirrhosus ehrhardti	SMF 11716	27.6	24.4	14.1	13.1	11.5	13.5		5.0	10.0	9.7	5.5	17.2	10.6
	SMF 11717	28.5	25.1	13.9	13.4	11.3	13.3		5.0	10.3	9.6	5.8	17.9	11.0
	SMF 11718	27.9	24.8	13.4	12.9	11.3	13.5		4.9	9.9	9.7	5.7	17.9	10.7
Chrotopterus														
Chrotopterus auritus australis	BMNH 1.3.11.1	36.6	31.7	17.4	10.2	17.8	14.3		6.3	13.2	12.0	7.5	23.6	14.9
Chrotopterus auritus guianae	BMNH 4.5.7.20	35.6		16.4	18.5	16.9	13.4		5.9	13.1	12.1	8.0	23.3	14.5
Vampyrus auritus	ZMB 3755	37.2												
GLOSSOPHAGINAE														
Glossophaga														
Monophyllus leachii	BMNH 42.8.17.17									6.7	5.6	4.1		7.1
Lonchophylla														
Lonchophylla mordax	BMNH 3.9.5.34	23.9	22.3	9.6	9.1				4.3	8.0	5.4	4.0	15.8	8.5
Lionycteris														
Lionycteris spurrelli	BMNH 13.8.10.1	18.9	17.6	8.8	8.0	7.3			4.3	6.0	4.5	3.1	12.1	6.2

APPENDIX 2.—*Continued.*

Taxon	Specimen	1	2	3	4	5	6	7	8	9	10	11	12	13
Anoura														
Anoura geoffroyi	BMNH 11a	25.6	24.1	14.0		10.0	9.8	8.8	5.1	9.6	6.1		17.8	10.0
Lonchoglossa wiedi aequatoris	NR 6	22.5	21.7	11.9		9.3	9.2	8.4	4.6	8.4	5.9		15.4	8.5
	NR 8	21.6	21.0	11.8	10.0	9.4	9.0	8.2	4.4	7.9		4.4	15.1	8.2
Scleronycteris														
Scleronycteris ega	BMNH 7.1.1.671						8.3		4.3	7.5	4.8	4.3	15.7	7.6
Lichonycteris														
Lichonycteris obscura	BMNH 95.4.27.1	19.6	18.7	10.9		8.3	8.3	7.6	4.2	6.3	4.6		12.9	6.5
Hylonycteris														
Hylonycteris underwoodi	BMNH 3.2.1.5	22.8	22.1	14.2		8.6	8.6	7.7	4.2	8.2		4.1	5.7	8.6
Platalina														
Platalina genovensium	BMNH 27.11.19.38	32.8				10.0	10.0		4.9	10.7	5.3	4.6	12.6	11.4
Choeroniscus														
Choeronycteris godmani	BMNH 79.12.24.1	19.4	18.4	11.6		7.7	7.9	7.1		6.8	3.7	3.1		
Choeronycteris inca	BMNH 12.9.5.2						8.5		3.9	4.7	7.7	3.9	7.8	7.8
Choeronycteris minor	SMNS 441					8.6	8.8	7.7	3.9	4.7	4.7		8.2	8.2
CAROLLIINAE														
Carollia														
Carollia verrucata	BMNH 106.a							7.7		7.0	8.1	4.8	14.0	7.7
Glossophaga amplexicauda	MNHN A.291			9.4					5.9	7.2	7.3	5.0	14.1	8.7
Phyllostoma bicolor	ZSM 126	23.3												
Vespertilio perspicillatus	BMNH 67.4.12.597									7.4	8.1	4.7	13.9	8.0
Rhinophylla														
Rhinophylla pumilio	ZMB 3060	19.5					8.1		5.5	5.3	6.4	4.8	12.7	6.8
Rhynophylla cumilis	SMNS 289(1)	18.7	16.9				8.1		5.4	5.2	6.5	4.5		
	SMNS 289(2)						7.8		5.2	5.0	5.8	4.3		
	ZMB 3346	18.2												
STENODERMINAE														
Sturnira														
Corvira bidens	BMNH 15.7.11.7	22.8	20.5	9.9		11.6	9.9		5.5	6.4	6.8	5.0	13.9	7.3
Phyllostoma albescens	NMW (not numbered)	22.4										6.2		
Phyllostoma excisum	ZSM 137	20.8												

APPENDIX 2.—Continued.

Taxon	Catalogue no.													
Phyllostoma lilium	MNHN (not numbered)	23.6		10.0			10.9		6.4	6.8	8.3	6.4	14.3	7.8
Sturnira spectrum	BMNH 42.12.2.4													
Uroderma														
Uroderma thomasi	BMNH 1.2.1.37	24.4	22.7		13.6	11.7	10.0		5.6	8.7	9.5	6.1	15.5	9.2
Vampyrops														
Artibeus vittatus	ZMB 568	31.4	28.9	16.4	18.9	15.5	12.7	12.9	7.2	12.8	14.5	8.1	16.4	13.6
Phyllostoma lineatum	MNHN 953			12.0			10.9		6.2	9.3	10.4	6.9	19.0	10.0
Vampyrops dorsalis	BMNH 99.12.5.1	27.5	25.5	14.5	15.3	13.4	11.0		6.4	11.0	11.6	5.7	15.7	11.7
Vampyrops lineatus sacrillus	BMNH 23.12.12.9	23.8		11.1	14.0		10.2			8.7	9.4	6.3	17.4	9.4
Vampyrops oratus	BMNH 14.7.27.1	26.6	24.2	12.7	14.9	12.7	11.1	11.0	6.2	10.0	11.2	6.4	15.3	11.7
Vampyrops recifinus	BMNH 81.2.16.4	24.1	21.7		14.0	11.7	10.3	10.7	5.6	8.9	10.5	5.6	13.0	9.5
Vampyrops zarhinus incarum	BMNH 12.1.15.1	20.4				10.2	8.7	9.3	5.2	6.8	7.7			7.5
Vampyrodes														
Vampyrodes ornatus	BMNH 24.3.1.63	27.7	24.7		16.7	13.5	11.2		6.2	10.2	12.0	6.9	17.4	
Vampyrodes caracciolae	BMNH 89.6.10.2			12.6					6.2	9.6	11.2	6.5	16.7	10.6
Vampyressa														
Chiroderma bidens	BMNH 69.3.31.12				11.6				6.2	8.3		4.8		7.2
Phyllostoma pusillum	ZSM 1843/2											4.1		
Vampyressa melissa	BMNH 26.5.3.4	21.6	20.2			10.5	9.1		5.0	7.0	9.7	5.3	13.7	7.5
Vampyressa nymphaea	BMNH 9.7.17.40				12.2	10.4	9.3		4.7	7.3	8.9			6.8
Vampyressa thyone	BMNH 97.11.7.77	19.1	17.4		11.0	9.6	8.4		5.0	6.2	7.9	4.8	11.2	6.8
Vampyressa venilla	BMNH 24.3.1.73	17.4	15.8		10.5	8.7	7.9		4.6	5.4	7.1	4.3	10.5	5.9
Chiroderma														
Chiroderma doriae	BMNH 49.8.16.29						11.8		6.2	10.6	13.2	6.6	18.9	11.5
Chiroderma salvini	BMNH 68.8.16.2	27.4	24.9		16.7				6.2	9.6	12.2	6.5	17.8	10.4
Chiroderma villosum	ZMB 408											6.3		
Mesophylla														
Mesophylla macconnelli	BMNH 1.6.4.64	17.8	15.8		10.2		7.8		4.5	6.0	6.9	4.0	10.4	6.6
Artibeus														
Artibeus carpolegus	BMNH 47.12.27.13	28.8	25.7		17.7	15.4	12.8		7.4	10.2	12.4	7.9	18.6	10.5
Artibeus cinereus bogatensis	BMNH 99.11.4.35	20.7							5.1	6.9	8.4	5.6	12.6	6.8
Artibeus concolor	ZMB 2617	22.0	19.4	9.5	13.4	11.8	10.0	10.6	5.6	7.5	9.3	6.0	13.2	7.8

APPENDIX 2.—Continued.

Taxon	Specimen												
Artibeus fallax	RNH 13083	30.7	27.4	15.6	18.1	16.2	12.9		7.5	11.1	13.7	8.7	20.5	12.0
	ZMB 566	30.9	27.4	15.0	18.9	16.2	13.0	12.9	7.2	11.3	14.2	8.7	19.9	12.2
Artibeus glaucus	BMNH 94.8.6.13	20.8	19.0		10.8		9.3		5.6	6.6	8.7	8.8	12.9	7.2
Artibeus jamaicensis aequatorialis	BMNH 0.2.9.13	29.3	26.7		18.2	16.0	13.0		7.1	11.2		8.8	19.8	12.2
Artibeus nanus	BMNH 89.1.30.5	17.6	15.2		11.3	9.5	8.6		4.5	5.5	8.0	8.1	10.6	5.9
Artibeus planirostris grenadensis	BMNH 96.11.8.6				17.1	15.2	12.2		7.0	10.4	8.0	8.1	18.6	11.3
Artibeus planirostris trinitatis	BMNH 97.6.7.1	27.9	24.7		16.7	15.0	12.2		7.0	9.8	12.2	7.6	18.1	10.3
Artibeus pumilio	BMNH 24.3.1.52	19.2	17.0		11.1	9.7	8.8		4.8	5.9	7.9	5.1	11.2	6.0
Artibeus (Dermanura) quadrivittatus	RNH 13114													
Artibeus (Dermanura) rosenbergi	BMNH 97.11.7.76	20.4	18.5		11.2	9.9	8.7		4.5	6.5	7.9	5.1		6.9
Artibeus turpis	BMNH 88.8.8.29	19.4	17.0		11.7	9.9	8.9		4.9	6.4	8.8	5.5	12.2	6.6
Artibeus watsoni	BMNH 0.7.11.19				11.6		8.6		4.8	6.2	8.3	5.5	12.3	7.0
Phyllostoma planirostre	ZSM 66	30.9				16.7	13.1		7.6	11.3	13.7	8.6	20.7	12.9
Enchisthenes														
Artibeus hartii	BMNH 92.9.7.8	20.5	18.6			10.6	9.2		6.1		8.1	5.4		7.6
Pygoderma														
Phyllostoma bilabiatum	ZSM 114	21.5												
Stenoderma (Pygoderma) microdon	ZMB 2713	20.0		6.6		12.2	10.4	11.3	7.6	5.5	7.3	6.0	10.8	5.4
Ametrida														
Ametrida centurio	BMNH 1957.a.	16.3	13.5		11.4	9.7	8.9		4.5	5.0	7.9	4.5	8.6	5.3
Centurio														
Centurio senex	BMNH (not numbered)	18.1	15.4		15.1	11.6			6.1	5.4	10.9	5.6	9.1	5.8
DESMODONTINAE														
Desmodus														
Desmodus murinus	ZSM 57	26.0												
	ZSM 59	25.2												
Edostoma cinerea	MNHN 958D				12.4	12.2	12.2		5.3			6.2	15.2	4.6
	MNHN 958G	25.4	22.6	9.4	12.8	13.1	13.2		5.9			6.4	14.8	4.1
Diaemus														
Desmodus youngii	RNH 12088	25.0	21.6		13.2	12.6	13.0		6.0		3.1	5.6	13.8	4.2
Diaemus youngi cypselinus	BMNH 28.7.21.64	26.3	23.0		14.3	13.5		14.0	6.5		3.5	6.2	15.2	4.2

APPENDIX 2.—Continued.

Taxon	Catalog No.												
Diphylla													
Diphylla centralis	BMNH 3.3.3.3	22.7	20.0	12.5		11.1	7.2	3.0			5.0		
NATALIDAE													
Natalus													
Natalus stramineus	BMNH (not numbered)	16.6	15.2	8.6	7.6	8.2	8.0	3.2	7.2	5.8	4.0		7.5
Phodotes tumidirostris continentis	BMNH 11.5.25.13	16.6	15.2	8.4	7.7	8.4	8.0	3.4	7.0	5.5	4.0	11.9	7.4
THYROPTERIDAE													
Thyroptera													
Thyroptera bicolor	RNH 17551	14.5	13.2		6.5	7.2		2.6	5.5		5.0		6.0
VESPERTILIONIDAE													
Myotis													
Myotis chiloensis alter	BMNH 0.6.29.23	16.5	15.3	10.0	8.1	7.8		4.0	6.0	6.2	4.0	11.2	6.5
Myotis dinellii	BMNH 0.7.9.4								5.6	3.6	10.5		6.0
Myotis lucifugus fortidens	BMNH 88.8.8.18	15.1	14.0	9.6	7.9	7.4		3.9	5.4	6.1	4.2	9.8	5.5
Myotis simus	BMNH 81.5.12.2	14.3	13.4		7.7	7.0		4.0	5.0	5.6	4.1		6.8
Myotis thysanodes aztecus	BMNH 58.6.2.3	13.6						4.1	6.3	6.7	4.1		5.3
Vespertilio arsinoe	RNH 17635				7.0	6.8		3.8	4.9	5.4	3.6	9.2	5.3
Vespertilio ferrugineus	RNH 17363							5.2	6.8	7.2	5.2		7.2
	RNH 17364	17.3	16.6		9.4	8.7		5.0	6.1	7.1	4.8		6.6
Vespertilio leucogaster	RNH 17622							5.0	5.3	5.9			6.6
Vespertilio levis	MNHN 864							3.9	5.6		3.5		5.6
Vespertilio nubilus	ZSM 121	15.2							5.0				
Vespertilio parvulus	RNH 17621												5.3
Vespertilio (Leuconoe) pilosus	MNHN (not numbered)	20.5	19.5	12.9	10.2	9.7		4.8	8.2	8.7	6.0	14.8	8.9
Pipistrellus													
Vespertilio isidori	MNHN 865					6.9		3.7	4.5		3.9		4.4
Vespertilio lacteus	RNH 17624								3.9				4.2
Eptesicus													
Eptesicus argentinus	BMNH 98.3.4.6	17.9	17.0		9.2	7.8		3.8	6.7	7.7	5.7	12.4	7.2
Eptesicus chiriquinus	BMNH 3.3.3.1	16.4		11.1	8.6	7.7		3.9	6.3	6.8	5.1	12.2	6.6
Eptesicus fidelis	BMNH 1.2.4.1	14.1	13.5	9.3	7.7	7.0		4.0	4.9	5.7	4.2	9.5	5.2
Eptesicus fuscus pelliceus	BMNH 98.7.1.28							4.3	7.4	6.3			8.2

APPENDIX 2.—*Continued.*

Taxon	Specimen												
Eptesicus inca	BMNH 94.8.6.1	16.8	16.3		8.9	8.0		4.4	6.3	7.0	5.1	11.8	6.8
Eptesicus montosus	BMNH 2.1.1.1	16.3	15.7		8.6	8.8		4.4	5.9	6.7	5.1	10.9	6.3
Eptesicus punicus	BMNH 99.8.1.1	14.6	13.9	8.8	7.4	6.9		3.6	5.4	5.9	4.2	9.9	5.7
Vespertilio arctoideus	ZSM 144										5.1		
Vespertilio fuscus peninsulae	BMNH 95.34.10.14	16.9	16.0	11.5	8.9	8.2		3.9	6.2	6.9	5.0	11.6	6.7
Vesperus melanopterus	RNH 12092								6.0	7.2	5.0		6.5
Nycticeius													
Vespertilio aenobarbus	RNH 17623							3.0	4.0	4.5	3.3		4.3
Rhogeessa													
Rhogeessa bombyx	BMNH 13.10.29.1	14.5	13.3	9.3	7.4	6.5		3.6	5.1	6.0	4.2	9.3	5.8
Rhogeessa io	BMNH 94.9.25.1							3.2	4.3	5.1	3.5		4.9
Rhogeessa velilla	BMNH 99.8.1.5	12.0	11.5		6.5	5.6		3.2	4.3	5.2	3.4	8.0	4.8
Lasiurus													
Atalapha egregia	ZMB 3762	17.3											
Atalapha frantzii	ZMB 2707	12.6											
	ZMB 2707	12.3											
	ZMB 3451	12.5											
	ZMB 595	16.8											
Atalapha pallescens													
Dasypterus ega argentinus	BMNH 98.3.4.9	15.8	15.0	10.8	8.5	8.6		4.7	5.6	7.3	5.9	10.7	6.3
Dasypterus ega fuscatus	BMNH 99.9.6.5							4.8	5.8	7.6	6.3	11.3	6.7
Dasypterus ega panamensis	BMNH 0.7.11.1			10.9		8.1		4.3	5.7	7.3	6.2	11.1	6.4
Dasypterus ega xanthinus	BMNH 98.3.1.14	16.6	15.5	11.0	8.9	8.2		4.7	5.7	7.6	6.2	11.1	6.4
Lasiurus caudatus	BMNH 44.10.9.7										6.3		
MOLOSSIDAE													
Molossops													
Dysopes abrasus	RNH 17374	25.6		16.2	15.8			5.6	7.5	9.3	5.4	13.8	8.3
Molossops mastivus	BMNH 10.11.10.3								8.7	10.2		16.6	9.9
Molossops temminckii sylvia	BMNH 98.3.4.21	14.7	14.2	8.9	8.7	7.2		3.7	5.7	6.6	3.8	9.7	6.1
Molossus cerastes	BMNH 1.8.1.13	22.5	21.4		14.8			5.2	8.4	10.3	6.7	15.2	9.5
Molossus (Molossops) planirostris	ZMB 2513	17.5	16.5	7.6	10.9	8.5	6.7	4.1	6.5	7.3	4.9	11.5	7.1
Molossus planirostris paranus	BMNH 1.7.11.15	18.0	17.2		11.5	8.4		4.6	6.5	7.7	5.0	12.0	7.2

APPENDIX 2.—*Continued.*

Tadarida												
Dysopes multispinosus ZMB 2614	17.3	16.0		9.8	9.3	8.5		6.3	7.9			6.8
ZMB 2614	*17.2*											
ZSM 134	*17.0*											
Dysopes naso												
Dysopes nasutus RNH 17576	17.3	16.3		10.1	9.5	8.7	4.3	6.3	7.5	4.4		
Nyctinomus brasiliensis MNHN 802								6.6	7.1	4.3		
MNHN 803								6.3	6.9	4.3		
Nyctinomus macrotis BMNH (not numbered)	22.4	21.1	9.2	11.6	10.9	9.8	4.0	8.7	8.3	4.9	15.4	9.3
Mormopterus												
Nyctinomus kalinowskii BMNH 94.8.6.7	14.8	13.7		8.6	8.0	7.2	3.2	4.8	5.8	3.5		5.2
Eumops												
Dysopes glaucinus NMW (not numbered)	*25.0*											
Dysopes longimanus ZSM 56	*22.5*						5.6					
ZSM 60	*24.7*											
Eumops dabbenei BMNH 14.4.4.8	32.2	30.0		19.0	16.1		6.0	13.1	13.2	8.1		14.4
Eumops delticus BMNH 23.8.9.7	18.8	17.9		10.9	10.3	9.1	4.2	7.2	7.9	4.7	12.7	7.9
Eumops geijskesi RNH 12943	21.4	19.4		12.5	10.8	10.2	4.1	8.4	8.9	5.1		9.0
Eumops patagonicus BMNH 23.12.12.18	17.8	16.4		10.5	9.9	9.0	4.2	6.7	7.5	4.1	11.6	7.2
Molossus maurus BMNH 1.6.4.34	21.8	20.2		12.4	10.7	9.7	4.0	8.3	8.8	5.2	14.7	9.1
Promops nanus BMNH 0.7.11.99						8.2	3.6	6.2	6.8	3.9	10.9	6.7
Promops trumbulli BMNH 99.11.2.1	28.9	27.3		14.2		12.3	5.0	11.1	10.9	6.5	20.4	12.3
Promops												
Molossus fosteri BMNH 1.8.1.17	19.0	17.4		11.3	11.0	9.8	4.1	6.9	8.2	4.6	11.8	7.8
Molossus nasutus ZSM 136	18.2					9.4	4.0	6.8	7.7	4.4	11.8	7.5
Promops ancilla BMNH 6.5.8.4	18.4	17.0		10.8	10.7	9.2	3.8	6.6	7.9	4.2	11.4	7.2
Promops centralis BMNH 94.2.5.4	21.5				12.0	10.4	4.3	8.3	9.7	5.3	14.2	9.3
Promops davisoni BMNH 21.5.21.1	20.0	18.3		11.7	11.2	9.7	3.8	7.5	8.7	4.8	12.7	8.2
Promops occultus BMNH 2.11.7.24	21.0	19.7		12.6	11.7	10.3	4.1	7.9	9.1	4.9	13.6	8.5
Molossus												
Dysopes alecto RNH 13023	23.1		8.2			11.2	4.6	8.3	10.3	6.3		9.4
Molossus burnesi BMNH 5.1.8.7	16.6	14.5			10.0	8.8	3.9	5.9	7.3	4.2	10.6	6.6
Molossus fluminensis BMNH (not numbered)	22.8	20.3		13.5	13.5	11.2	4.7	8.5	10.2	5.8	15.1	9.5

APPENDIX 2.—Continued.

Molossus longicaudatus	MNHN 792	15.6	13.9		9.5	8.4	3.2	5.8	7.3	4.0		6.2
Molossus obscurus	MNHN A.419/225	17.8	15.8		10.8	8.9	3.4	6.3	7.8	4.5	11.2	7.2
Molossus obscurus currentium	BMNH 98.3.4.28	19.9	18.1		10.6	8.9	3.8	7.0	8.4	4.9		7.8
Molossus rufus	MNHN A.428/224	24.5	20.8	13.8	13.3	11.0	4.4	8.3	9.8	6.2	15.4	9.3
Molossus tropidorhynchus	BMNH 54.12.27.6	16.6	15.0		9.9	8.3	3.4	6.1	7.4	4.2	10.6	6.8

APPENDIX 3.—Index to species-group names arranged alphabetically by author. Species-group names are followed by page number.

APPENDIX 3.—*Continued.*

JENTINK

Desmodus youngii, 68
Vesperus melanopterus, 80

JOHNSON

Molossus milleri, 98

KAPPLER

Rhynophylla cumilis, 48

KERR

Vespertilio molossus minor, 99

LATASTE

Molossus fluminensis, 96

LEACH

Aello cuvieri, 28
Artibeus jamaicensis, 61
Madataeus lewisii, 64

LICHTENSTEIN AND PETERS

Hyonycteris discifera, 69

LINNAEUS

Vespertilio leporinus, 25
Vespertilio perspicillatus, 48

LÖNNBERG

Lonchoglossa wiedi aequatoris, 41

MILLER

Myropteryx pullus, 19
Promops nanus, 93

MILLER AND G. M. ALLEN

Myotis chiloensis alter, 70
Myotis lucifugus fortidens, 70
Myotis thysanodes aztecus, 71

PETERS

Artibeus concolor, 60
Artibeus fallax, 60
Artibeus quadrivittatus, 63
Artibeus vittatus, 52

Atalapha egregia, 82
Atalapha frantzii, 82
Atalapha pallescens, 82
Balantiopteryx plicata, 22
Chiroderma villosum, 58
Choeronycteris minor, 45
Diclidurus scutatus, 23
Dysopes gigas, 91
Glossonycteris lasiopyga, 40
Lophostoma brasiliense, 32
Molossus ferox, 92
Molossus planirostris, 85
Mormops megalophylla, 28
Nyctinomus musculus, 90
Peropteryx kappleri, 20
Phylloderma stenops, 36
Rhinophylla pumilio, 49
Schizostoma behnii, 30
Schizostoma hirsutum, 31
Sphaeronycteris toxophyllum, 66
Stenoderma microdon, 66
Uroderma bilobatum, 51
Vampyrops helleri, 53
Vampyrus auritus, 37
Vespertilio pilosus, 75

SAUSSURE

Carollia azteca, 46
Molossus mexicanus, 88

SCHINZ

Vespertilio calcaratus, 21
Vespertilio leucogaster, 73

SPIX

Glossophaga amplexicaudata, 37
Molossus nasutus, 94
Noctilio albiventer, 25
Noctilio rufus, 25
Phyllostoma planirostre, 64
Proboscidea saxatilis, 16
Vampyrus bidens, 33

TEMMINCK

Dysopes abrasus, 84
Dysopes alecto, 96
Dysopes nasutus, 88
Emballonura lineata, 15

APPENDIX 3.—*Continued.*

Urocryptus bilineatus, 18
Vespertilio aenobarbus, 80
Vespertilio arsinoe, 71
Vespertilio ferrugineus, 72
Vespertilio lacteus, 76
Vespertilio parvulus, 74

THOMAS

Anthorhina peruana, 34
Anthorhina picata, 34
Artibeus glaucus, 61
Artibeus hartii, 64
Artibeus pumilio, 62
Artibeus rosenbergi, 63
Artibeus watsoni, 64
Balantiopteryx io, 21
Centronycteris centralis, 21
Chilonycteris davyi fulvus, 26
Chiroderma doriae, 58
Choeronycteris godmani, 44
Choeronycteris inca, 45
Chrotopterus auritus australis, 36
Chrotopterus auritus guianae, 37
Corvira bidens, 49
Cyttarops alecto, 24
Dasypterus ega argentinus, 83
Dasypterus ega fuscatus, 83
Dasypterus ega panamensis, 83
Dasypterus ega xanthinus, 83
Diaemus youngi cypselinus, 68
Diclidurus virgo, 24
Diphylla centralis, 69
Dirias irex, 24
Eptesicus argentinus, 77
Eptesicus chiriquinus, 77
Eptesicus fidelis, 77
Eptesicus fuscus pelliceus, 77
Eptesicus inca, 78
Eptesicus montosus, 78
Eptesicus punicus, 78
Eumops dabbenei, 91
Eumops delticus, 92
Eumops patagonicus, 92
Glyphonycteris sylvestris, 29
Hylonycteris underwoodi, 43
Lichonycteris obscura, 42
Lionycteris spurrelli, 39
Lonchophylla mordax, 39
Mesophylla macconnelli, 59
Molossops mastivus, 85

Molossops temminckii sylvia, 85
Molossus burnesi, 96
Molossus cerastes, 85
Molossus fosteri, 94
Molossus maurus, 93
Molossus obscurus currentium, 98
Molossus planirostris paranus, 85
Myotis dinellii, 70
Myotis simus, 71
Nyctinomus kalinowskii, 90
Peronymus cyclops, 20
Phodotes tumidirostris continentis, 69
Phyllostoma latifolium, 35
Platalina genovensium, 44
Promops ancilla, 94
Promops centralis, 94
Promops davisoni, 95
Promops occultus, 95
Promops trumbulli, 93
Rhogeessa alleni, 81
Rhogeessa bombyx, 81
Rhogeessa io, 81
Rhogeessa velilla, 81
Saccopteryx bilineata centralis, 16
Saccopteryx canescens, 17
Saccopteryx gymnura, 17
Saccopteryx infusca, 22
Saccopteryx pumila, 17
Scleronycteris ega, 41
Tonatia laephotis, 33
Vampyressa melissa, 56
Vampyressa nymphaea, 57
Vampyressa thyone, 57
Vampyressa venilla, 58
Vampyrodes ornatus, 54
Vampyrops caracciolae, 55
Vampryops dorsalis, 53
Vampyrops lineatus sacrillus, 53
Vampyrops oratus, 54
Vampyrops recifinus, 54
Vampyrops zarhinus incarum, 54
Vespertilio fuscus peninsulae, 79

TOMES

Lasiurus caudatus, 84
Lonchorhina aurita, 31

TSCHUDI

Phyllostoma oporophilum, 51

APPENDIX 3.—*Continued.*

WAGNER

Chilonycteris gymnonotus, 26
Chilonycteris personata, 27
Chilonycteris rubiginosa, 28
Desmodus murinus, 67
Dysopes albus, 95
Dysopes auritus, 86
Dysopes glaucinus, 90
Dysopes gracilis, 87
Dysopes longimanus, 91
Dysopes naso, 87
Emballonura brevirostris, 18
Emballonura insignis, 16
Emballonura macrotis, 20
Phyllostoma albescens, 49
Phyllostoma ambylotis, 32
Phyllostoma bicolor, 47
Phyllostoma bilabiatum, 65

Phyllostoma calcaratum, 47
Phyllostoma discolor, 35
Phyllostoma excisum, 50
Phyllostoma fumarium, 50
Phyllostoma pusillum, 56
Phyllostoma scrobiculatum, 30
Vespertilio arctoideus, 79
Vespertilio carbonarius, 72
Vespertilio nubilus, 74
Vespertilio splendidus, 75

WATERHOUSE

Desmodus dorbignyi, 67
Vespertilio chiloensis, 72

WIED—NEUWIED

Diclidurus albus, 22
Diclidurus freyreisii, 23

APPENDIX 4.—Index to species-group names arranged alphabetically by museum. Species-group names are followed by page number in italics.

APPENDIX 4.—*Continued.*

Myotis thysanodes aztecus Miller and
   G. M. Allen, 1928, *71*

Natalus stramineus Gray, 1838, *69*

Nyctinomus kalinowskii Thomas, 1893, *90*

Nyctinomus macrotis Gray, 1839, *89*

Nyctinomus megalotis Dobson, 1876, *90*

Peronymus cyclops Thomas, 1924, *20*

Phodotes tumidirostris continentis
   Thomas, 1911, *69*

Phyllophora megalotis Gray, 1842, *29*

Phyllostoma childreni Gray, 1838, *32*

Phyllostoma elongata Gray, 1842, *30*

Phyllostoma latifolium Thomas, 1901, *35*

Phyllostoma soricinum É. Geoffroy
   St.-Hilaire, 1910, *38*

Platalina genovensium Thomas, 1928, *44*

Promops ancilla Thomas, 1915, *94*

Promops centralis Thomas, 1915, *94*

Promops davisoni Thomas, 1921, *95*

Promops nanus Miller, 1900, *93*

Promops occultus Thomas, 1915, *95*

Promops trumbulli Thomas, 1901, *93*

Pteronotus davyi Gray, 1838, *28*

Rhinops minor Gray, 1866, *48*

Rhogeessa alleni Thomas, 1892, *81*

Rhogeessa bombyx Thomas, 1913, *81*

Rhogeessa io Thomas, 1903, *81*

Rhogeessa velilla Thomas, 1903, *81*

Saccopteryx bilineata centralis Thomas,
   1904, *16*

Saccopteryx canescens Thomas, 1901, *17*

Saccopteryx gymnura Thomas, 1901, *17*

Saccopteryx infusca Thomas, 1897, *22*

Saccopteryx pumila Thomas, 1914, *17*

Scleronycteris ega Thomas, 1912, *41*

Scotophilus macleayii Gray, 1843, *79*

Sturnira spectrum Gray, 1842, *51*

Tonatia laephotis Thomas, 1910, *33*

Uroderma thomasi Andersen, 1906, *52*

Vampyressa melissa Thomas, 1926, *56*

Vampyressa nymphaea Thomas, 1909, *57*

Vampyressa thyone Thomas, 1909, *57*

Vampyressa venilla Thomas, 1924, *58*

Vampyrodes ornatus Thomas, 1924, *54*

Vampyrops caracciolae Thomas, 1889, *55*

Vampyrops dorsalis Thomas, 1900, *53*

Vampyrops lineatus sacrillus Thomas,
   1924, *53*

Vampyrops oratus Thomas, 1914, *54*

Vampyrops recifinus Thomas, 1901, *54*

Vampyrops zarhinus incarum Thomas,
   1912, *54*

Vespertilio chiloensis Waterhouse, 1838, *72*

Vespertilio fuscus peninsulae Thomas,
   1898, *79*

Vespertilio leporinus Linnaeus, 1758, *25*

Vespertilio perspicillatus Linnaeus, 1758, *48*

Vesperugo dorianus Dobson, 1885, 77

MUSEO NACIONAL DE CIENCIAS NATURALES,
MADRID

Noctilio zaparo Cabrerra, 1907, *25*

MUSÉUM NATIONAL D'HISTOIRE NATURELLE,
PARIS

Diclidurus scutatus Peters, 1869, *23*

Edostoma cinerea d'Orbigny, 1835, *68*

Glossophaga amplexicauda É. Geoffroy
   St.-Hilaire, 1818, *46*

Glossophaga caudifer É. Geoffroy St.-Hilaire,
   1818, *41*

Molossus longicaudatus É. Geoffroy
   St.-Hilaire, 1805, *97*

Molossus obscurus É. Geoffroy St.-Hilaire,
   1805, *98*

Molossus rufus É. Geoffroy St.-Hilaire,
   1805, *99*

Molossus rugosus d'Orbigny, 1837, *89*

Nyctinomus brasiliensis I. Geoffroy
   St.-Hilaire, 1824, *89*

Phyllostoma crenulatum É. Geoffroy
   St.-Hilaire, 1810, *34*

Phyllostoma elongatum É. Geoffroy
   St.-Hilaire, 1810, *35*

Phyllostoma lilium É. Geoffroy St.-Hilaire,
   1810, *50*

Phyllostoma lineatum É. Geoffroy St.-
   Hilaire, 1810, *52*

Phyllostoma soricinum É. Geoffroy
   St.-Hilaire, 1810, *38*

Schizostoma brachyote Dobson, 1879, *31*

Schizostoma hirsutum Peters, 1869, *31*

Stenoderma rufa Desmarest, 1820, *65*

Vespertilio isidori d'Orbigny and Gervais,
   1847, *76*

Vespertilio levis I. Geoffroy St.-Hilaire,
   1824, *74*

Vespertilio molossus minor Kerr, 1792, *99*

Vespertilio pilosus Peters, 1869, *75*

APPENDIX 4.—*Continued.*

Vespertilio polythrix I. Geoffroy St.-
 Hilaire, 1824, *75*

NATURHISTORISCHES MUSEUM WIEN, VIENNA

Chilonycteris gymnonotus Wagner,
 1843, *26*
Chilonycteris personata Wagner, 1843, *27*
Dysopes albus Wagner, 1843, *95*
Dysopes auritus Wagner, 1843, *86*
Dysopes glaucinus Wagner, 1843, *90*
Emballonura brevirostris Wagner, 1843, *18*
Emballonura insignis Wagner, 1855, *16*
Emballonura macrotis Wagner, 1843, *20*
Phyllostoma ambylotis Wagner, 1843, *32*
Phyllostoma scrobiculatum Wagner,
 1855, *30*

NATURHISTORISKA RIKSMUSEUM, STOCKHOLM

Lonchoglossa wiedi aequatoris Lönnberg,
 1921, *41*

RIJKSMUSEUM VAN NATUURLIJKE HISTORIE,
 LEIDEN

Artibeus fallax Peters, 1865, *60*
Artibeus quadrivittatus Peters, 1865, *63*
Desmodus youngii Jentink, 1893, *68*
Dysopes abrasus Temminck, 1827, *84*
Dysopes alecto Temminck, 1827, *96*
Dysopes nasutus Temminck, 1827, *88*
Emballonura lineata Temminck, 1840, *15*
Eumops geijskesi Husson, 1962, *92*
Myotis surinamensis Husson, 1962, *71*
Phylloderma stenops Peters, 1865, *36*
Proboscidea saxatilis Spix, 1823, *16*
Thyroptera bicolor Cantraine, 1845, *70*
Urocryptus bilineatus Temminck, 1838, *18*
Vampyressa nattereri Goodwin, 1963, *57*
Vespertilio aenobarbus Temminck,
 1840, *80*
Vespertilio arsinoe Temminck, 1840, *71*
Vespertilio ferrugineus Temminck, 1840, *72*
Vespertilio lacteus Temminck, 1840, *76*
Vespertilio leucogaster Schinz, 1821, *73*
Vespertilio parvulus Temminck, 1840, *74*
Vesperus melanopterus Jentink, 1904, *80*

NATUR-MUSEUM UND FORSCHUNGS-INSTITUT
 SENCKENBERG, FRANKFURT A. M.

Trachops cirrhosus ehrhardti Felten, 1956, *36*

STAATLICHES MUSEUM FÜR NATURKUNDE IN
 STUTTGART, STUTTGART

Choeronycteris minor Peters, 1868, *45*
Rhynophylla cumilis Kappler, 1881, *49*

ZOOLOGISCHES MUSEUM DER HUMBOLDT-
 UNIVERSITÄT ZU BERLIN, BERLIN

Artibeus concolor Peters, 1865, *60*
Artibeus fallax Peters, 1865, *60*
Artibeus vittatus Peters, 1860, *52*
Atalapha egregia Peters, 1871, *82*
Atalapha frantzii Peters, 1871, *82*
Atalapha pallescens Peters, 1871, *82*
Balantiopteryx plicata Peters, 1867, *22*
Carollia azteca Saussure, 1860, *46*
Chiroderma villosum Peters, 1860, *58*
Diclidurus albus Wied-Neuwied, 1826, *22*
Diclidurus freyreisii Wied-Neuwied,
 1821, *23*
Dysopes gigas Peters, 1864, *91*
Dysopes gracilis Wagner, 1843, *87*
Dysopes multispinosus Burmeister, 1861, *87*
Dysopes naso Wagner, 1840, *87*
Dysopes temminckii Burmeister, 1854, *84*
Glossonycteris lasiopyga Peters, 1868, *40*
Hyonycteris discifera Lichtenstein and
 Peters, 1855, *69*
Molossus ferox Peters, 1861, *92*
Molossus mexicanus Saussure, 1860, *88*
Molossus planirostris Peters, 1865, *85*
Mormops megalophylla Peters, 1864, *28*
Myropteryx pullus Miller, 1906, *19*
Nyctinomus musculus Peters, 1861, *90*
Peropteryx kappleri Peters, 1867, *20*
Proboscidea saxatilis Spix, 1823, *16*
Rhinophylla pumilio Peters, 1865, *48*
Rhynophylla cumilis Kappler, 1881, *49*
Schizostoma behnii Peters, 1866, *30*
Sphaeronycteris toxophyllum Peters, 1882, *66*
Stenoderma microdon Peters, 1863, *66*
Uroderma bilobatum Peters, 1866, *51*
Vampyrops helleri Peters, 1866, *53*
Vampyrus auritus Peters, 1856, *37*
Vespertilio calcaratus Schinz, 1821, *21*
Vespertilio hilarii I. Geoffroy St.-Hilaire,
 1824, *80*

ZOOLOGISCHES STAATS-SAMMLUNG
 MÜNCHEN, MUNICH

Chilonycteris rubiginosa Wagner, 1843, *28*

APPENDIX 4.—*Continued.*

Desmodus murinus Wagner, 1840, *67*
Dysopes gracilis Wagner, 1843, *87*
Dysopes longimanus Wagner, 1843, *91*
Dysopes naso Wagner, 1840, *87*
Glossophaga amplexicaudata Spix, 1823, *37*
Molossus nasutus Spix, 1823, *94*
Noctilio albiventer Spix, 1823, *25*
Noctilio rufus Spix, 1823, *25*
Phyllostoma albescens Wagner, 1847, *49*
Phyllostoma bicolor Wagner, 1840, *47*
Phyllostoma bilabiatum Wagner, 1843, *65*
Phyllostoma calcaratum Wagner, 1843, *47*

Phyllostoma discolor Wagner, 1843, *35*
Phyllostoma excisum Wagner, 1842, *50*
Phyllostoma fumarium Wagner, 1847, *50*
Phyllostoma oporophilum Tschudi, 1844, *51*
Phyllostoma planirostre Spix, 1823, *64*
Phyllostoma pusillum Wagner, 1843, *56*
Proboscidea saxatilis Spix, 1823, *16*
Vampyrus bidens Spix, 1823, *33*
Vespertilio arctoideus Wagner, 1855, *79*
Vespertilio carbonarius Wagner, n.d., *72*
Vespertilio nubilus Wagner, 1855, *74*
Vespertilio splendidus Wagner, 1845, *75*

APPENDIX 5.—List of specimens used for comparative purposes and referred to in text by catalogue number and museum prefix TCWC (Texas Cooperative Wildlife Collection, Department of Wildlife and Fisheries Sciences, Texas A&M University). Specimens are listed in numerical order.

7303	♂	*Hylonycteris underwoodi*	México: Oaxaca: 1½ mi. N San José Chacalapa
9831	♀	*Lichonycteris obscura*	Nicaragua: Bluefields: 6 mi. W Rama, 50 ft.
9932	♂	*Carollia castanea*	Costa Rica: Puntarenas: 9 mi. ENE Puerto Golfito
10734	♀	*Phyllostomus discolor*	Honduras: Tegucigalpa: 12 mi. N Tegucigalpa 2800 ft.
11169	♀	*Choeroniscus godmani*	México: Veracruz: 2 km. SE Sontecomapán
11702	♂	*Tonatia silvicola*	Ecuador: Loja: 15 mi. N Catacocha, 2000 ft.
11705	♀	*Tonatia silvicola*	Perú: Piura: 4 mi. W Suyo
11717	♂	*Phyllostomus discolor*	Ecuador: El Oro: Portovelo, 2800 ft., 1½ mi. S Zaruma
11718	♂	*Phyllostomus discolor*	Ecuador: El Oro: Portovelo, 2800 ft., 1½ mi. S Zaruma
11725	♂	*Phyllostomus discolor*	Ecuador: El Oro: Portovelo, 2800 ft. 1½ mi. S Zaruma
11744	♀	*Phyllostomus discolor*	Ecuador: El Oro: Portovelo, 2800 ft. 1½ mi. S Zaruma
11866	♀	*Glossophaga soricina*	Colombia: Bolivar: 5 mi. E Sincelejo
11883	♂	*Anoura caudifer*	Ecuador: Napo Pastaza: 8 mi. WNW Puyo, 3800 ft.
11886	♀	*Anoura caudifer*	Ecuador: Napo Pastaza: 8 mi. WNW Puyo, 3800 ft.
11900	♀	*Lionycteris spurrelli*	Perú: San Martin: Previsto, 1500 ft., 25 km. W Aguaytia
12036	♀	*Carollia perspicillata*	Perú: Huánuco: 2 mi. N Tingo María
12068	♀	*Carollia brevicauda*	Ecuador: Napo Pastaza: 8 mi. WNW Puyo
12090	♀	*Carollia brevicauda*	Perú: Huánuco: 19 mi. S Tingo María
12172	♂	*Vampyrops dorsalis*	Ecuador: El Oro: Mina Miranda, 1 mi. N Zaruma, 4600 ft.
12281	♀	*Artibeus glaucus*	Ecuador: Napo Pastaza: 8 mi. WNW Puyo, 3800 ft.
12282	♂	*Artibeus jamaicensis*	Ecuador: Loja: 15 mi. N Catacocha
12319	♀	*Artibeus lituratus*	Colombia: Popayán, Cauca
12327	♀	*Artibeus planirostris*	Perú: Loreto: 11 mi. SE Pucallpa
12334	♀	*Artibeus planirostris*	Perú: Loreto: 11 mi. SE Pucallpa
12338	♂	*Artibeus cinereus*	Perú: Loreto: 61 mi. SE Pucallpa, 500 ft.
12481	♂	*Molossus* sp.	Ecuador: Napo Pastaza: Shell-Mera, 3800 ft., 4 mi., W Puyo
12576	♂	*Molossus* sp.	Perú: Huánuco: 19 mi. S Tingo María, 2800 ft.
12585	♂	*Molossus* sp.	Ecuador: El Oro: Portovelo, 2800 ft., 1½ mi. S Zaruma
12682	♂	*Myotis albescens*	Perú: Huánuco: 19 mi. S Tingo María, 2800 ft.
12702	♂	*Myotis nigricans*	Ecuador: El Oro: 9 mi. S Zaruma, 2000 ft.
12703	♀	*Myotis chiloensis*	Ecuador: Carchi: Gruta Rumichaca, 2 mi. E La Paz, 8700 ft.
12710	♂	*Myotis oxyotus*	Perú: Huánuco: 9 mi. S Huánuco, 7200 ft.
12713	♀	*Myotis nigricans*	Colombia: Valle: 6 mi. N Palmira
12715	♂	*Myotis nigricans*	Perú: Huánuco: 19 mi. S Tingo María, 2800 ft.
12727	♂	*Myotis albescens*	Perú: Piura: 4 mi. W Suyo, 1000 ft.
12737	♂	*Myotis nigricans*	Perú: Piura: 4 mi. W Suyo, 1000 ft.
12838	♂	*Promops* sp.	Perú: Piura: Suyo 1000 ft.
14553	♂	*Diclidurus virgo*	Honduras: 12 mi. N San Pedro Sula
16412	♂	*Phylloderma stenops*	México: Chiapas: 10 mi. W Mal Paso, 400 ft.
16424	♂	*Carollia perspicillata*	México: Chiapas: 16 mi. NW Palenque, 100 ft.
16460	♀	*Carollia subrufa*	México: Oaxaca: 4 mi. E Tapanatepec
16501	♂	*Carollia brevicauda*	México: Chiapas: 16 mi. NW Palenque, 100 ft.

APPENDIX 6.—List of species-group names for which no type specimens were found. Entries consist of species-group name, author, and year of publication. Abbreviated museum names are included, following year of publication, when we were reasonably certain of the actual or most likely repository. These specimens are presumed lost, unless otherwise noted, although some might yet exist. However, they were not identifiable as primary types in 1966 (or 1976 for British Museum).

Alectops ater Gray, 1866, BMNH

Anoura wiedii Peters, 1869, MNHN

Arctibeus leucomus Gray, 1848, BMNH[1]

Atalapha cineria brasiliensis Pira, 1905

Celaeno brocksiana Leach, 1821, D. Brookes Museum[2]

Centurio mexicanus Saussure, 1860, MNHN[3]

Chilonycteris osburnii Tomes, 1861, BMNH[4]

Choeronycteris mexicana Tschudi, 1844, ZMB

Choeronycteris peruana Tschudi, 1844, ZMB

Depanycteris isabella Thomas, 1920, BMNH[5]

Dermanura cineria Gervais, 1855, MNHN

Diphylla ecaudata Spix, 1823, ZSM[6]

Desmodus fuscus Burmeister, 1854, ZMB[7]

Desmodus mordax Burmeister, 1879, ZMB[7]

Dysopes amplexicaudatus Wagner, 1850, NMW[8]

Dysopes caecus Rengger, 1830

Dysopes holosericeus Wagner, 1843, NMW[8]

Dysopes leucopleura Wagner, 1843 NMW[8]

Dysopes olivaceofuscus Wagner, 1850, NMW[8]

Dysopes rufocastaneus Schinz, 1844, ZMB

Dysopes thyropterus Schinz, 1844, ZMB

Emballonura brunnea Geravis, 1855, MNHN

Furia horrens F. Cuvier, 1828, MNHN

Furipterus caerulescens Tomes, 1856, Tomes Collection[4]

Glossophaga ecaudata É. Geoffroy St.-Hilaire, 1818, MNHN

Hyonycteris albiventer Tomes, 1856, Tomes Collection[4]

Ischnoglossa nivalis Saussure, 1860, MNHN[3]

Lasiurus borealis salinae Thomas, 1902, BMNH

Lasiurus grayi Tomes, 1857, BMNH

Lophostoma silvicolum d'Orbigny, 1835, MNHN

Macrotus mexicanus Saussure, 1860, MNHN[3]

Molossops aequatorianus Cabrera, 1917, MNCN[9]

Molossus acuticaudatus Desmarest, 1820, MNHN

Molossus aztecus Saussure, 1860, MNHN[3]

Molossus (Molossops) brachymeles Peters, 1865, ZSM

Molossus crassicaudatus É. Geoffroy St.-Hilaire, 1805[10]

Molossus fumarius Spix, 1823, ZSM

Molossus laticaudatus É. Geoffroy St.-Hilaire, 1805[10]

Molossus moxensis d'Orbigny, 1837, MNHN

Molossus myosurus Tschudi, 1844, ZSM

Molossus ursinus Spix, 1823, ZSM

Myotis nigricans osculatii Cabrera, 1917, MNCN[9]

Myotis thomasi Cabrera, 1901, MNCN[9]

Noctilio affinis d'Orbigny, 1837, MNHN

Noctilio dorsatus Desmarest, 1818, BMNH[11]

Noctilio rufipes d'Orbigny, 1835, MNHN

Noctilio unicolor Desmarest, 1818[10]

Noctilio vittaus Schinz, 1821, ZMB

Nycticejus ega Gervais, 1856, MNHN

Nycticejus varius Pöppig, 1835[12]

Nyctiplanus rotundatus Gray, 1849 (for 1848)

Peropteryx leucoptera Peters, 1867, ZMB[13]

Phyllodia parnellii Gray, 1843, BMNH[14]

Phyllostoma angusticeps Gervais, 1855, MNHN

Phyllostoma bennettii Gray, 1838, BMNH

Phyllostoma bernicaudum Schinz, 1821[15]

Phyllostoma brachyotum Schinz, 1821

Phyllostoma chrysocomos Wagner, 1855[16]

Phyllostoma erythromos Tschudi, 1844

Phyllostoma innominatum Tschudi, 1844

Phyllostoma longifolium Wagner, 1843, NMW[8]

Phyllostoma macrophyllum Schinz, 1821, ZMB[17]

Phyllostoma maximus Wied-Neuwied, 1821

Appendix 6.—*Continued.*

Phyllostoma obscurum Schinz, 1821, ZMB

Phyllostoma rotundum É. Geoffroy St.-Hilaire, 1810, MNHN[18]

Phyllostoma spiculatum Illiger, *in* Lichtenstein, 1825, ZMB

Phyllostoma superciliatum Schinz, 1821, ZMB

Proboscidea rivalis Spix, 1823, ZSM

Promops bonariensis Peters, 1874

Rhinolophus ecaudatus Schinz, 1821, ZMB

Schizostoma minutum Gervais, 1855, MNHN

Spectrellum macrurum Gervais, 1856 MNHN

Stenoderma tolteca Saussure, 1860, MNHN[3]

Sturnira spectrum Gray, 1842, BMNH

Thyroptera tricolor Spix, 1823, ZSM

Trachops fuliginosus Gray, 1857

Tylostoma mexicana Saussure, 1850, MNHN[3]

Vampyrus auricularis Saussure, 1860, MNHN[3]

Vampyrus cirrhosus Spix, 1823, ZSM

Vespertilio albescens É. Geoffroy St.-Hilaire, 1806, MNHN

Vespertilio auripendulus Shaw, 1800

Vespertilio borealis Müller, 1776

Vespertilio brasiliensis Desmarest, 1819

Vespertilio brasiliensis Spix, 1823, ZSM

Vespertilio caninus Schinz, 1821, ZMB

Vespertilio cinnamomeus Wagner, 1855, ZSM

Vespertilio derasus Burmeister, 1854, ZMB

Vespertilio espadae Cabrera, 1901, MNCN[9]

Vespertilio furinalis d'Orbigny, 1847, MNHN

Vespertilio guianensis Lacepede, 1789

Vespertilio hastatus Pallas, 1767

Vespertilio hypothrix d'Orbigny and Gervais, 1847, MNHN

Vespertilio innoxius Gervais, 1841, MNHN

Vespertilio (Myotis) kinnamon Gervais, 1856 (for 1855), MNHN

Vespertilio labialis Kerr, 1792

Vespertilio leptura Schreber, 1774

Vespertilio maximiliani Fischer, 1829

Vespertilio maximus É. Geoffroy St.-Hilaire, 1806, MNHN

Vespertilio minor Fermin, 1765

Vespertilio molossus Pallas, 1766 (part)[19]

Vespertilio nasutus Shaw, 1800

Vespertilio nigricans Schinz, 1821, ZMB[20]

Vespertilio nitens Wagner, 1855

Vespertilio noveboracensis Erxleben, 1777

Vespertilio oxyotus Peters, 1866

Vespertilio ruber É. Geoffroy St.-Hilaire, 1806, MNHN

Vespertilio soricinus Pallas, 1766

Vespertilio spectrum Linnaeus, 1758

Vespertilio spixii Fischer, 1829[21]

Vespertilio splendidus Wagner, 1845, ZSM

Vespertilio subflavus F. Cuvier, 1832, MNHN

Vespertilio villosissimus É. Geoffroy St.-Hilaire, 1806, MNHN

---

[1] The holotype of *A. leucomus*, and only specimen of *Pygoderma bilabiatum* in the British Museum known to Dobson (1878:537), might be the single skin and skull from Santa Catarina, Brazil.

[2] At least some of the Brookes Museum specimens are now in the British Museum.

[3] De Saussure's specimens were housed in the Comparative Anatomy Laboratory of the Paris museum and, for the most part, were dissected by Saussure. However, duplicates were exchanged with other museums, perhaps especially the Berlin museum, and it would appear that any remaining syntypes of Saussure names should be looked for elsewhere.

[4] The Tomes Collection went to the British Museum in 1907, but some of the bat specimens once in his collection could not be traced. It should be noted that R. F. Tomes acquired many of the bats when the Zoological Society of London disposed of its collections in 1855, most of the material going directly to the British Museum. Tomes names based on specimens in the Zoological Society of London Collection that were acquired by the British Museum in 1855 and reported by Dobson (1878) are identified as BMNH specimens; those assigned in this list to the Tomes Collection could not be traced in the British Museum.

[5] This specimen (20.7.14.24) was reported to be in alcohol, skull removed.

[6] *Glossophaga diphylla* Fischer, 1829, appears to be a junior objective synonym.

[7] All of Burmeister's specimens are assumed to have been in the Berlin museum inasmuch as the only ones found were there.

[8] From Johann C. Natterer's collection. Most of Natterer's specimens were returned to the Vienna museum, although Wagner kept a few duplicates. A number of specimens prepared by Johann Natterer but now without identifying labels are extant at Vienna.

APPENDIX 6.—*Continued.*

[9]Personnel at the Museo Nacional de ciencias Naturales were of the opinion that Cabrera had taken his type specimens to Argentina. Of the several Cabrera names, only a paratype for *Noctilio zaparo* (this catalogue, p. 25) could be found at the Madrid museum.

[10]From Azara's Paraguayan collection, none of which appears to have been at the Paris museum.

[11]This name appears to be a junior objective synonym of *Vespertilio leporinus* Linnaeus, 1758, by way of renaming *Pteropus leporinus*, Erxleben, 1777. An account for V. *leporinus* is included in this catalogue, p. 25.

[12]*Nycticeus poeppigii* Lesson, 1836, appears to be a junior objective synonym.

[13]From Kappler's Surinam collection.

[14]Originally based on two specimens from Jamaica, one was listed by Dobson (1878:452).

[15]Spelling error for *Phyllostoma brevicaudum* (see Pine, 1972:29). According to Avila-Pires (1965:6), a syntype, no. 1333, American Museum of Natural History, could not be located; another syntype, RNH 17692, was not examined by us, but Pine (1972:30) reported that K. F. Koopman examined this specimen.

[16]Appears to be a renaming of *Sturnira spectrum* Gray, 1842.

[17]*Macrophyllum nieuwiedii* Gray, 1838, is a junior objective synonym.

[18]*Desmodus rufus* Wied-Neuwied, 1824, appears to be a junior objective synonym although Wied-Neuwied (1826) reported one or more specimens from Fazenda Muribeca on the Rio Itabapoana, separating the states of Espírito Santo and Rio de Janeiro, Brazil.

[19]*Vespertilio molossus major* Kerr, 1792, and *Molossus fusciventer* É. Geoffroy St.-Hilaire, 1805, are junior objective synonyms of *Vespertilio molossus* Pallas, as fixed by Husson, 1962:257 (see accounts for *Vespertilio molossus minor* Kerr, 1792, and *Molossus longicaudatus* É. Geoffroy St.-Hilaire, 1805, pp. 99 and 97, respectively, in this catalogue).

[20]LaVal (1973:9) designated a neotype (36877, Los Angeles County Museum) for *Vespertilio nigricans* Schinz.

[21]We suspect that this name is a junior objective synonym of *Vespertilio brasiliensis* Spix, 1823.